The Surprising *Archaea*

THE SURPRISING *ARCHAEA*

Discovering Another
Domain of Life

John L. Howland

New York   Oxford

OXFORD UNIVERSITY PRESS

2000

Oxford University Press

Oxford   New York
Athens   Auckland   Bangkok   Bogotá   Buenos Aires   Calcutta
Cape Town   Chennai   Dar es Salaam   Delhi   Florence   Hong Kong   Istanbul
Karachi   Kuala Lumpur   Madrid   Melbourne   Mexico City   Mumbai
Nairobi   Paris   São Paulo   Singapore   Taipei   Tokyo   Toronto   Warsaw

and associated companies in
Berlin   Ibadan

Library of Congress Cataloging-in-Publication Data
Howland, John L.
  The surprising *archaea* : Discovering another
  domain of life / by John L. Howland.
    p.   cm.
  Includes index.
  ISBN 0-19-511183-4
  1. Archaebacteria.   I. Title.
QR82.A69H69 2000
579.3'21—dc21   99-27106

9 8 7 6 5 4 3 2 1

Printed in the United States of America
on acid-free paper

Preface

This book records the archaeal rise from obscurity—the group was entirely unknown two decades ago—to their current prominent place in molecular and evolutionary biology. I hope that the book may also function as antidote to blank expressions that flitted across various faces when this writing project was first bruited about: one's friends obviously needed some educating in such matters. In short, this book is a work of advocacy and, to be truthful, it is one with an even larger agenda than promoting *Archaea*, as marvelous as they may be. Thus, my hope is also to promote wider awareness of the rich world of microbes in general, a world that we tend to ignore until it crosses our path in the form of disease or other unpleasantness. Finally, the last element of the plan is to employ the story of the *Archaea* as a vehicle for communicating something of the attractiveness and utility of current molecular biology. This goal is almost unavoidable: molecular biological techniques were central to the very discovery of the *Archaea*. Indeed, the procedures of molecular biology that have been used to define and study *Archaea* are themselves of such elegance that it would have been a great shame to omit them. (I didn't.)

It is my recurring privilege to teach seminars on "the origin of life and its early evolution" to undergraduates; this book originated in these courses. Indeed, its real genesis was in questions asked by the students in those courses and, to the extent that this book answers some of those questions, I hope that some of my former students read it, even without the certainty of a high grade. I would dedicate the book to them—the former students—but, as you will see, I have someone else in mind. But I should record here that several recent undergraduates, including Mason Bragg, Chris Molvar, and Marissa Zahler, read and commented on portions of the manuscript, for which I thank them very much. I am also blessed with exceedingly helpful colleagues and am particularly indebted to Bill Steinhart, Tom Settlemire, David Page, and Dan Levine for engaging in useful discussions, serving as edifying examples, or both. My closest colleague, Cynthia, provided her customary moral support, while retaining perspective and her sense of humor throughout the entire project. It should also be recorded that the exemplary patience and good sense of Kirk Jensen qualifies him for sainthood, or perhaps to be executive editor at Oxford University Press.

I am extremely grateful to Professor Karl Stetter for his encouragement, which took concrete form in a number of electron micrographs and other photographs. In that connection, I also thank his colleague, Dr. Reinhard Rachel, for his careful preparation and transmission of the images used. I am also grateful to Professor John Leigh for electron micrographs of *Methanococcus jannaschii* (a leading character in this story). Finally, thanks are due to David Page for preparation of drawings of membrane lipids and Kevin Johanen for his digital wizardry leading to several other of the illustrations.

I first learned about the great enjoyment associated with studying the microbial world from the late William R. Sistrom. So great was his influence on me that microbial biochemistry has been the only scientific game that I have ever really enjoyed playing, although I have tried my hand at some others. Most of what I know about microbes I learned, one way or another, under Bill's influence and I gratefully dedicate this book to him.

Islesboro, Maine   J. L. H.
May 1999

Foreword

On November 3, 1977, an article entitled "Scientists Discover a Form of Life that Predates Higher Organisms" appeared on the front page of the *New York Times.* The article began:

> Scientists studying the evolution of primitive organisms reported today the existence of a separate form of life that is hard to find in nature. They described it as a "third kingdom" of living material, composed of ancestral cells that abhor oxygen, digest carbon dioxide and produce methane.

This discovery was considered important enough to be the subject of an additional article in the following Sunday's *Times,* with the title "A New Form of Life?" It was also deemed sufficiently newsworthy to eclipse a major press release that same week by the National Academy of Sciences. (The National Academy had announced the first cloning of the human growth hormone gene, an impressive milestone in the application of molecular biology to medicine).

With the hindsight of two decades, the significance of discovering this "New Form of Life" holds up rather well: the *Times* was amply justified in its front-page treatment of the story. The group of microorganisms described in the article, which are now called *Archaea*, were immensely interesting, and their discovery could not have been anticipated in the way that many discoveries are. There were very few prior hints of their existence, as a group, before this 1977 bombshell. The *Archaea* also achieved rapid notoriety by turning out to inhabit some of the most forbidding and unexpected places on Earth. They were discovered living in volcanic vents on the seafloor, in highly concentrated brines, in coal refuse piles heated by spontaneous combustion, in pockets of water in the Earth's crust at extremely high temperature and pressure, and even within the cytoplasm of giant amoebas. The *Archaea* also appear to be early products of evolution that had evolved relatively slowly during some billions of years, owing largely to the long-term stability of their rigorous environments.

Because they branched so long ago from other life, they constitute one of only three fundamental categories (or "domains") of organisms. The impact of their discovery was stunning: suddenly in 1977, biologists became aware of a third kind of life, where previously only two were known. Thus, prior to 1977, all organisms were considered either prokaryotes (bacteria) or eukaryotes (everything else, including plants, animals, and fungi). After that date, the tree of life was seen to have, not two, but three branches and the exact placement of the new one became the focus of research and debate among evolutionary biologists. It was soon concluded that the *Archaea* were extremely ancient, so that they appear able to shed light on evolution during very early chapters of life's history.

The discovery of the *Archaea*, with their exotic attributes and their evolutionary implications, came as a considerable surprise, and that tone of unexpectedness was evident in the press accounts of the time. Clearly, something truly interesting had occurred. This book explores that discovery and its significance, especially with regard to the place of the *Archaea* in the evolutionary scheme of things, as well as the diverse roles that they presently play in the biosphere. Thus, this book constitutes something of a biography of the *Archaea* as a group, comprising an account of the recent and fruitful unfolding of archaeal biology.

Contents

The Surprising *Archaea*

Meeting Some *Archaea*

A Formal Introduction

> Above all, introduce your main characters
> right away. Give the reader a chance to see if
> he likes them. It's only fair.
>
> Lawrence Durrell, *Reflections on a*
> *Marine Venus.*

Our characters are microorganisms belonging to the domain *Archaea*, and they are well worth meeting. Many are spectacularly eccentric in their lifestyle, living in scalding spring water, in the digestive systems of animals, or within the very cells of host organisms. A large number of them are fiendishly inventive in their ability to do exotic chemistry, carrying out reactions quite unknown in other organisms. Moreover, some are bizarre, to say the least, in their physical appearance, consisting of intricate networks, amorphous blobs, triangles, or postage-stamp-like squares. Although all of them are microscopic, their collective role in the world ecosystem is pervasive. And they all have a great deal to tell us about very early evolution and even, it appears, about our own evolutionary origins. Because a full appreciation of the *Archaea* will require excursions into molecular biology and evolutionary theory, it seems a good idea to start by getting to know several of them on a more personal, nonmolecular level.

We begin with introductions to three representative characters in the archaeal drama. They are interesting creatures in their own right, but also useful in illustrating the three major lifestyles that biologists have observed among the *Archaea*: thermophilic, methanogenic, and halophilic. These are ecological categories that represent overall patterns of adaptation to the environment and provide the framework for much of our thought about archaeal biology. Most *Archaea* belong to at least one of these categories. A number belong to two.

The thermophiles live at high temperatures, often inhabiting terrestrial or marine hot springs. As a rough guide, organisms that live at temperatures exceeding 50 degrees Centigrade (122 degrees Fahrenheit) are considered thermophilic. Here our representative of the thermophile category is *Sulfolobus*.

Methanogens can't tolerate oxygen and produce the gas methane; these organisms characteristically live in anaerobic muds or in the digestive systems of animals, and are represented here by *Methanococcus*.

Finally, halophiles live at salt concentrations high enough to kill most organisms. They occur in hypersaline bodies of water such as the Dead Sea or the Great Salt Lake. The halophiles are represented in this chapter by *Natronobacterium*.

These three representatives illustrate many important features of *Archaea*, so that they constitute an apt introduction to the group. These particular microorganisms have also been so extensively studied that they crop up repeatedly in any discussion of archaeal biology. We will encounter them again in various contexts, so that it will help to become a bit familiar with them here.

Before considering examples of the three great ecological categories of *Archaea*, it is necessary to issue something of a disclaimer: there is reason to believe that microorganisms representing additional archaeal lifestyles may even now be waiting in the wings to be discovered—*Archaea* that do not produce methane, cannot endure high salt concentrations, and are not thermophilic. Thus, it seems likely that we will soon be aware of additional lifestyle categories, some of them perhaps as different from "ordinary life" as the first three are. There is already evidence, in the form of anomalous nucleotide sequences recovered from the environment, that such additional categories do exist, but we do not yet have any real clues about their way of life. In all such matters, it is obviously desirable to maintain a flexible attitude, embracing new developments as they emerge. One can only anticipate with pleasure the discovery of new major forms of microbial life:it will be as if one had unexpectedly stumbled on the entire plant kingdom, having previously been quite unaware of its existence.

Clearly one should not lose sight of the newness of our experience with the *Archaea*; they were discovered quite unexpectedly, scarcely two decades ago, as a completely novel, but very old, branch of life—one of only three main branches. This monumental discovery, based on molecular biological observations that will be described in the next chapter, revealed a new form of life so different from known forms that we are still in the early stages of sorting things out. The newness of this situation leads archaeal biologists to consider themselves uncommonly fortunate at the present moment—to be engaged in an activity that is singularly full of promise. It likewise follows that one should not be surprised that our knowledge of the *Archaea* and their lifestyle categories may be somewhat fluid. But, for the time being, scientists are still only certain of the three categories, and it is now appropriate to meet examples of each of them, beginning with a rather famous thermophile.

Introducing *Sulfolobus acidocaldarius*

Envision a hot spring situated in Yellowstone National Park in northwestern Wyoming. Its hot water bubbles up gently, emitting a whitish vapor that smells faintly of rotten eggs. The rim of the spring, devoid of vegetation, is colored reddish brown by minerals, and, as it turns out, also by algae and an extensive mix of other microorganisms. The Yellowstone landscape is enlivened by an assortment of thermal springs, steam vents (or "fumaroles"), hot mud pools, and the region's premier attractions, dozens of geysers, the most spectacular of which send periodic plumes of hot water thirty meters into the air. The water that emerges from these features is often close to the boiling point at atmospheric pressure, 100 degrees Centigrade.*

But most geothermal features are less dramatic and we focus our attention on a quiet, evenly flowing spring. Its temperature is about 80 degrees: too hot to touch, but definitely not boiling. And although hot springs and other geothermal systems may seem quite exotic—and the Yellowstone Basin is assuredly a special place—geothermal activity is actually quite widespread on the Earth's surface, occurring at numerous geologically active sites on all continents, Antarctica included.

Yellowstone springs were often given colorful names by early explorers of the region. Local geological shapes, water color, and the distinctive forms of vapor plumes have led to identifying such geothermal features as Octopus Spring, Sapphire Spring, the Punch Bowl, Steamboat Spring, and the Beehive. Some of these, such as Octopus Spring, are famous among scientists interested in hot-spring biology, being sites of extensive research and the source of many important thermophilic organisms. Their amusing names thus take on an added scientific luster.

The chemical content of hot springs affects their ability to support life and influences the selection of organisms that can inhabit them. Some springs, like the Mammoth Hot Spring cluster, are rich in calcium, whose insoluble salts form huge elephantine deposits. Others contain enough particulate sulfur to line themselves with a bright yellow crust; still others contain iron, and so stain the surrounding rocks pink or rust. Also, although the water from many springs and geysers is neutral in pH—neither particularly acidic nor particularly basic—some springs are exceedingly acidic. In such cases, their pH value can approximate the acidity of vinegar or human stomach liquid. But, whereas the stomach contains hydrochloric acid, these springs are sour with sulfuric acid. In such situations, sulfuric acid can serve as a chemical precursor of hydrogen sulfide, H_2S, the gas that endows many springs and fumaroles with their characteristic aroma of rotten eggs.

In view of such formidable temperatures and such corrosive acidity, an innocent bystander might imagine these springs to be devoid of life. Indeed, hot springs and geyser pools consti-

*All temperatures henceforth are given in the centegrade scale.

tute a notable danger to wildlife: bison, deer, and wolves have been known to perish after accidental immersion. Nor are tourists or biologists immune to the effects of the scalding waters. Thus, such waters are singularly unsuited for the support of most life—indeed they contain no fish, plants, or other higher organisms. But, strange as it may seem, they teem with microbial life, a great deal of it archaeal.

The first hint that boiling springs were anything but sterile came from observation of unicellular algae in their warm outflows. And as early as 1966, Thomas Brock, a microbiologist whose name is closely associated with the study of thermophilic bacteria, found pink bacteria growing in Mushroom and Octopus Springs at temperatures near 70 degrees Centigrade. Subsequently, a yellow-pigmented bacterial sample from Mushroom Spring provided the source for isolation of *Thermus aquaticus*,* one of the first thermophilic eubacteria to receive extensive study. This was the first organism found to grow at a temperature greater than 70 degrees. Although hot-spring microbes can live freely suspended in the spring water, more commonly they live in gelatinous mats. These mats are complex ecosystems several millimeters thick containing photosynthetic *Eubacteria* on the exterior surface, where they can receive light, and nonphotosynthetic *Archaea* and *Eubacteria* beneath.

Thomas Brock has been a pioneer in the study of thermophiles—according to many, he is the "father" of thermophilic microbiology. Even before 1964, he and his wife, Louise Brock, had discovered algae living in Yellowstone springs at over 60 degrees and bacteria at near 80. The Brocks had clearly not been inhibited by then-prevailing dogma that life was impossible at much above 50 degrees, so that the whole world of thermophiles became open to them. A compelling account of their microbial adventures is found in his essay, "The Road to Yellowstone—and beyond," cited in the Additional Reading section at the end of the book.

Information about microbial life in hot springs has been obtained by suspending microscope slides directly into the

*A reminder about scientific names: they come in pairs. The first specifies the genus (plural: genera), the second, the species. A particular genus can include several different species (e.g., *Sulfolobus acidocaldarius, Sulfolobus solfataricus,* and *Sulfolobus shibatae*).

springs. This immersion slide technique is widely used by microbial ecologists. When slides had remained only one day in Yellowstone springs that were near the boiling point, bacteria were found adhering to the slides and could be stained, viewed through a microscope, and photographed. Using this approach, growth was observed from many boiling pools often the growth was so extensive that a film could be seen on the slide with the unaided eye. So if one examines water from any of these springs, one is very likely to find diverse single-celled organisms. And if one examines water from hotter and more acidic springs, one has a good chance of encountering a microbe called *Sulfolobus acidocaldarius*, an organism with worldwide distribution, living in comparable springs in volcanically active regions of Iceland, Italy, and New Zealand.

Sulfolobus acidocaldarius is an archaeon, a member of the domain *Archaea*. We will soon say a great deal more about domains, but here it is enough to define them as the most inclusive division of living things. Thus, all organisms belong to one of three domains: the *Archaea*, the *Eubacteria*, and the *Eukarya*.

Because *Sulfolobus* flourishes in hot, acidic water, it is called both thermophilic and acidophilic. It is unicellular, an isolated cell about two micrometers (i.e., two one-millionths of a meter) long. Its cells are irregular and lumpy in shape, being roughly spherical, and decorated with variable lobes. Moreover, the name is instructive. The generic (first) name suggests the lobed shape and a predilection for sulfur; the specific (second) name specifies its fondness for acidic environments and its affinity for vulcanism, caldera being volcanic basins.

Although some of its archaeal relatives possess flagella and are capable of swimming about, *Sulfolobus* is quite immotile. Many of its closest relatives are killed by oxygen, yet *Sulfolobus* can live in either the presence or absence of air. When the oxygen from air is not available for respiration, this adaptable organism makes do with sulfur or sulfur compounds instead. In effect, under such conditions, *Sulfolobus* can "breathe" sulfur.

Scientists have been aware of *Sulfolobus acidocaldarius* since the early 1970s, even prior to the discovery of the *Archaea* as a group. *Sulfolobus* had been initially judged to be a bacterium that was remarkably adapted for life under hot and acidic conditions, but an ordinary bacterium nonetheless. But it turned out not to be a bacterium at all: it was virtually a founding member of the new *Archaea*. Indeed, as the validity of *Archaea* as a distinct domain became widely accepted, *Sulfolobus acidocaldarius* came to be considered by many as the quintessential representative of the group as a whole, a sort of archaeal icon.

By 1998, over 400 scientific publications focused on *Sulfolobus acidocaldarius* and its close relatives. Many of these publications concentrated on the organism's remarkable resistance to elevated temperature while others described the genetic system of the creature, a single circular DNA molecule, or the manner in which genetic information is expressed, or "read," in the process of protein synthesis. Indeed, the scientific study of *Sulfolobus*, and other members of its tribe, extends beyond the cells of these creatures in the sense that pieces of their genetic material have now been incorporated into other organisms. For example, *Sulfolobus* genes have been transferred to cells of that microbial workhorse, *Escherichia coli*, and gene expression and the character of the gene products examined under the more familiar conditions possible there. This is very ecumenical, amounting to transfer of genetic material all the way from one domain to another.

The current and persistent level of interest in *Sulfolobus* is illustrated by the decision of a group of Canadian molecular biologists to sequence its entire genome—its total complement of genetic material (DNA). This consists of about three million nucleotide base pairs. By contrast, the entire human genome contains about three billion. Also by way of comparison, the first archaeal genome to be sequenced was that of *Methanococcus jannaschii*, with a mere 1.66 million base pairs.

Introducing a Methanogen: *Methanococcus jannaschii*

The archaeon *Methanococcus jannaschii* is noteworthy for reasons that go far beyond the elucidation of its genome nucleotide sequence, having a life, "worthy," as they say at funerals, "of celebration." It was discovered by means of a research submarine at a location where geothermally heated water issues from vents in the seabed to mix with the icy water of the ocean. Cells of *Methanococcus jannaschii* are spherical and about one micrometer in diameter. The cells also sport a variable number of flagella and so are able to swim (or, more accurately, tumble) about in their hot, watery environment.

The name, *Methanococcus jannaschii*, is also instructive. A coccus is any roughly spherical microorganism and the "methano" part identifies this beast as a methanogen—a synthesizer of the natural gas methane. The specific (second) part of the name honors Holger Jannasch, a pioneer in the discovery and exploration of hydrothermal vents and the leader of the expedition that obtained the vent sample from which this organism was first isolated. A genus usually includes a number of closely related species: species of *Methanococcus* include *igneus, infernus, thermolithotrophicus, voltae,* and, of course, *jannaschii.* The first three names celebrate a thermophilic way of life; the last two honor discoverers—Alessandro Volta was an early observer of biological methane production.

Many *Archaea* are methanogens, the first members of that group to have been recognized as fundamentally distinct from bacteria. Many methanogens are thermophilic—denizens of hot places—such as *M. jannaschii.* All methanogens can live without oxygen, and many (including *M. jannaschii*) are seriously harmed by its presence. Methanogens can also exhibit odd nutritional requirements with, for example, growth of *M. jannaschii* being stimulated by the trace metals selenium and tungsten, both presumably serving as enzyme cofactors in its rather specialized metabolism. Other methanogens inhabit the bottom sediments of swamps, rice paddies, ponds, and streams. Still others are found in the digestive systems of animals.

Methanococcus jannaschii hails from undersea hydrothermal vents, and many of its attributes reflect this origin. Vent water is usually anaerobic—devoid of oxygen—and *M. jannaschii* is a strict anaerobe. Consistent with its lack of oxygen, vent water is also normally rich in chemically reduced compounds, compounds able to donate electrons to other chemicals.* Such chemically reduced compounds are required for methanoge-

---

*In chemistry, when a reduced compound donates electrons, this compound becomes oxidized, and the compound accepting electrons becomes reduced. Notice that reduction is always associated with oxidation and vice versa as addition of an electron to one atom depends on its removal from another. Note that oxygen is the chemical antithesis of a reduced compound: it is an avid electron acceptor and thus an extremely poor donor.

nesis to occur. Finally, vents are hot and *M. jannaschii* is understandably thermophilic, with optimum growth at about 85 degrees Centigrade and the ability to tolerate temperatures near 91 degrees. (This is 196 degrees Fahrenheit.) This archaeon can also live under the very high pressures found at the ocean floor, sometimes in excess of 200 atmospheres, one atmosphere being the normal pressure of air at sea level.

Its Discovery

In 1982, the research submersible *Alvin* dove to a depth of almost three kilometers at a geologically active region in the eastern Pacific Ocean. There in the dark, its bionic arm plucked a sample of material from the base of a vent "chimney" and the sample was subsequently sent to the University of Illinois for analysis. From it, J. A. Leigh was able to isolate an organism that now bears the name of *Methanococcus jannaschii*. Its original home had been a piece of vent chimney, a complex community containing a variety of bacterial and archaeal species. The ecology of such communities is based on nutrients in the vent effluents. These communities are quite unique in the living world: most ecosystems count photosynthesis as their ultimate energy source. Energy in these submarine systems, where sunlight does not penetrate, originates in microbial transformations of vent-water chemicals—in particular, chemicals in the reduced (electron-rich) state. An example of a reduced chemical that sustains growth of *Methanococcus* is hydrogen gas (H_2), which is a starting material for its methane production. Such chemical reactions provide energy for the cells that carry them out and those cells, or their products, are, in turn, utilized by the other resident organisms. Thus, the ecosystem is energetically isolated from solar radiation, which, in any case, fails to penetrate to such depths.

The vent where *M. jannaschii* was discovered was a "white smoker," so called because of the whitish turbidity that results from mineral precipitation when the hot vent water merges with the cold seawater. There are also "black smokers" whose inky plumes of effluent are colored by metallic sulfide minerals that come out of solution when their hot vent water is rapidly cooled. The vent chimneys are of considerable architectural interest themselves, sometimes appearing as elongated cones two or three meters high, composed of minerals precipitated from the vent water. In general appearance, they might be compared to the giant nests of African termites. When the *Alvin* nosed around among the clustered vents, its headlamps disclosed small forests of the chimneys. These proved to harbor a wide variety of thermophilic *Archaea*, including, of course, *Methanococcus jannaschii.*

A Pharaonic Microbe (*Natronobacterium pharaonis*)

The Wadi Al Natrun is a depression in the Western Desert about sixty miles northwest of Cairo. "Wadi" is Arabic for watercourse (with or without water) and, in this instance, a number of dry valleys lead from the fringes of the desert into the depression, where a cluster of ponds is located. The ponds are also fed by underground water seeping from the Nile River via a series of grassy swamps. These ponds are extremely saline because of evaporation and, owing to the chemical composition of the surrounding minerals, the water contains a high concentration of sodium carbonate. The carbonate makes the water extremely alkaline, with a pH of over ten. In contrast, the seepage swamps contain water at a neutral pH—around seven—as evaporation has not yet had a chance to concentrate the carbonate.

An ancient word for sodium carbonate is "natron," from which the depression obviously receives its name. Natron also figures in the process of mummification as practiced historically in the Nile Valley. To emphasize this connection with the name *Natronobacterium,* we turn to Herodotus's contemporary account of the use of natron:

> Then the body is placed in natrum for seventy days, and covered entirely over. After the expiration of that space of time, which must not be exceeded, the body is washed. (Herodotus, the Histories, Book Two)

The function of the "natrum" (natron) was largely to serve as a dehydrating agent—an essential part of mummification is removal of water from the tissues of the corpse.

Natronobacterium pharaonis is an archaeon originally isolated from Natrun. As its generic name suggests, it grows well with sodium carbonate in its growth medium. Also, it is evident that it was named at a time when a distinction between Archaea and true bacteria had not been established. Its specific (second) name is a bit redundant, but is redolent of ancient Egypt and reminds us that the pharaohs that physically inhabit museums today do so because their bodies spent the requisite seventy days soaking in a natron (sodium-carbonate, mummification-grade) bath.

This organism, *N. pharaonis*, is an example of the third great ecological category of archaeal microorganisms, the halophiles. The word denotes salt loving. Such organisms are able to grow in extremely concentrated salt solutions that would prove fatal to most other creatures. Interestingly, the ability to grow in concentrated salt solutions reflects, at least in part, resistance to the dehydrating effects of the salt. In other words, a salt, such as natron, is harmful to most organisms for the same reason that it is useful for mummification: it extracts water. And halophiles have evolved mechanisms that enable them to compensate for that water loss.

Recall that the first two archaeal groups that we encountered were the thermophiles and the methanogens and that it is possible for a particular organism to belong to both. Likewise, we will see that there are overlaps between halophiles and the other sorts of Archaea as well. Thus, these ecological categories, although widely employed and convenient in describing *Archaea*, are not rigid. Indeed, they are not taxonomic categories at all because they do not, in fact, represent evolutionary history: they simply reflect lifestyle.

The archaeon *N. pharaonis* is an ordinary rod-shaped microbe that lives under conditions that would kill most organisms. Not only does it grow in brine containing concentrated sodium chloride, but it also favors extremely alkaline conditions. In the laboratory, this halophile is maintained at a pH close to ten that is just as extreme and corrosive to most cells as the pH of two, in which *Sulfolobus* and other acidophilic Archaea are able to thrive.

In fact, these halophiles defend themselves against high salt concentrations, and consequent dehydration, by pumping salts across their external membrane. Their external membrane contains a protein that functions by absorbing light and using its energy to pump negatively charged ions, such as chloride or nitrate, across that membrane. The light-absorbing protein halorhodopsin is named for its striking chemical resemblance to the visual rhodopsin of the vertebrate eye.

This outward pumping of chloride makes sense: the archaeon lives in a high concentration of chloride, much higher than that of its interior. Therefore, the cell continuously uses light energy to extrude the chloride, thereby lowering the internal concentration. And when the negatively charged chloride ions exit from the cell, positively charged sodium ions follow passively, further aiding in the maintenance of a dilute interior. The ability of the halorhodopsin pump to transport nitrate as well as chloride is a bit enigmatic, especially as nitrate pumping is often more rapid than that of chloride. The problem is this: nitrate is not a major component of the salty ponds in the Wadi E1 Natrun and it is not clear why the halorhodopsin of *N. pharaonis* should have evolved the ability to transport it. Perhaps nitrate bears a molecular resemblance to some other anion that the protein really evolved to transport, but we don't know if that is the case or what that other anion might be—a nice mystery awaiting a solution.

The Wealth of Archaeal Species

This chapter has introduced the *Archaea* by confining itself to one example of each of their three major ecological groups. These three creatures are the representatives of many, and perhaps it is fitting here to tell how many, giving also a sense of the relative size of the three groups. The halophilic Archaea that have, thus far, been grown and studied in laboratories comprise nine genera and over twenty-five species. The methanogens include twenty-one genera and more than sixty-five species. Finally, those thermophilic Archaea that are not also methanogenic comprise seventeen genera and more than sixty species. And all these numbers will probably grow as the study of the *Archaea* proceeds.

One can see that methanogens enjoy a slight edge in diversity, probably owing to the greater variety of their ecological situations. Although they are restricted by their requirement for anaerobiosis, they are extremely widespread in the biosphere, occurring in a wide range of habitats, including the deep sea, freshwater bogs, and guts of cattle.

Other Players in the Archaeal Drama

The rest of this book describes a plethora of archaeal species, delving into some of the more intimate details of their lives as it goes along. As a rule, we will meet them one by one as appropriate contexts unfold, but it seems right to end this chapter with some vignettes that illustrate a little more of the variety of archaeal adaptation (beyond heat and salt tolerance and methane exhalation). In this fashion, the reader will obtain a preliminary sense of the scope of the archaeal world beyond the mere tabulation of numbers of species previously presented.

For example, a group of microbiologists carried out deep drilling into ancient soil about 190 meters below the surface in Hanford, Washington. Although this soil had not been near the surface for a considerable span of geological time—some seven million years—and the drillers took extensive precautions to avoid contamination from surface microorganisms, they obtained evidence for the presence of 746 bacterial and 190 archaeal organisms. This evidence was in the form of nucleotide sequences: the microbiologists did not necessarily obtain the living organisms themselves. However, we will learn that such sequences can be reliably linked to actual organisms and so are unambiguous indicators of their presence. In this way, it is possible to inventory an entire ecosystem, "populating" it with a collection of organisms, many of which are known only through their nucleotide signatures.

From the Hanford bacterial sequences, most were identifiable as ordinary soil-dwellers, but, surprisingly, some resembled sequences from a photosynthetic bacterium, *Chloroflexus*, of a sort found in the Sargasso Sea. Obviously, photosynthesis does not occur at a depth of 190 meters and, anyway, Washington state is a long way from the central Atlantic, so this resemblance is somewhat perplexing. The archaeal sequences were enigmatic as well: many most closely resembled those of thermo-

philic *Archaea*, particularly some that had previously been isolated from the Obsidian Pool, an extremely hot spring in
Yellowstone National Park. The deep soil of Hanford, Washington, certainly does not resemble a hot spring actually exhibits a temperature of about 17 degrees. One is thus faced
with strong evidence for the occurrence of "inappropriate"
microorganisms in this location. Similar findings in other areas
of archaeal ecology indicate that *Archaea* are distributed much
more widely than previously thought. The best place to search
for archaeal thermophiles may still be a hot spring or submarine vent, but they may also crop up just about anywhere else.

Indeed, they may even lurk at depths much greater than the
190 meters of the Hanford study: exploratory drilling deep into
the Earth's crust has increasingly turned up populations of
thermophilic *Archaea*, leading to the remarkable proposal of a
deep ecosystem of global extent. This would consist of large
populations of microorganisms, mostly archaeal, remote from
photosynthesis and subsisting by coupling their metabolism to
geochemical reactions. One can think of them as feeding, so
to speak, on rock. Not surprisingly, hints of these subterranean
populations have come from oil drilling, which has turned up
a variety of *Archaea* with attributes that reflect their rocky home.
For instance, deep pockets in the saline oil fields of Alsace,
France, are graced with spherical (coccoid) *Archaea* that are
both halophilic and methanogenic, but not especially thermophilic, growing optimally at about 38 degrees.

Calculations that attempt to estimate the global mass of such
cells that exist deep in the Earth's crust must take into account
the extremely large volumes of water occurring in innumerable
seams and pockets, the high densities of cells that have been
observed through drilling, and the large geographical area involved. There is clearly a very large biomass to consider, and
much of it must be archaeal.

We will examine this notion of this deep biosphere in some
detail later, after we have assembled more background information. The potential occurrence of immeasurable tons of *Archaea* growing in hot, dark seams in the global crust challenges
prevailing views of how the ecosphere is organized, both with
respect to the primacy of photosynthesis and the relative ecological contribution of the *Archaea* to life on Earth.

At the same time, this idea has stimulated speculation on the part of "exobiologists," who propose that life may have been transported throughout the universe—possibly even to our Earth—packaged in the depths of rocky, asteroid-like, celestial objects. Indeed, the same deep archaeal ecosphere notion is also being used to promote the view that life may, even now, be evolving independently in the rocky depths of planets other than our own or, perhaps, deep in the interior of certain of their satellites.

Finally, *Archaea* also turn up in the intestines of many varieties of animals, as in the case of large and complex populations of methanogens residing in cattle. Moreover, so many marine animals, both invertebrate and vertebrate, contain archaeal populations in their digestive systems that their feces may contribute significantly to the widespread distribution of free archaeal cells in seawater. Thus, *Archaea* have been observed (often as represented by their nucleotide sequences) living in the cavities of sponges and the guts of sea cucumbers, flounder, and mullet. But they do not necessarily inhabit other marine creatures, such as mussels and copepod crustaceans, indicating that the archaeal associations are species-specific and that they do not just arrive there, for example, by filter feeding. The microorganisms that inhabit fish guts appear related, as revealed by their nucleotide sequences, to free-living archaeal residents of seawater. Some appear closely connected to the methanogens and, in fact, methane synthesis has been observed directly in the intestines of codfish. Thus, one can see that *Archaea*, formerly considered residents of only extreme habitats, have a disconcerting way of cropping up very nearly everywhere in the biosphere—in deep rocks, the open sea, and animal intestines. This ubiquitous distribution will be a recurring theme throughout the remainder of this book.

Their Discovery

Discovering microorganisms is not what it used to be. When van Leeuwenhoek, in the seventh century, first described bacteria, he drew what he had observed through a microscope of his own manufacture: his landmark discovery of the bacteria was pure visual description. By the nineteenth century, chemistry had entered the discovery process. Pasteur, Lister, and, later, Beijerinck and Winogradsky described and classified new microorganisms on the basis of their nutrition and their metabolic capabilities. But, in the twentieth century, for reasons we will be considering, such structural and chemical criteria for identifying new microbes, or groups of microbes, diminished in their usefulness, and scientists increasingly turned to the newly developed methods of molecular biology. This transition has been difficult for biologists who lean toward natural history of the descriptive sort, but it has also been enormously productive. The discovery of the *Archaea* was perhaps the first great product of the new, molecular-biological approach to describing organisms and an account of the discovery is also an account of the transition to that new methodology.

Turning to the *Archaea*, we have already seen that they come in a variety of shapes. They are also a mixed lot in that many of them require oxygen, whereas some are killed by it. Some *Archaea* grow at elevated temperatures, but many don't. Some live in volcanically heated water and others in the digestive tract of cattle. A number of them can swim using flagella, whereas others are as motionless as a brick. One must wonder what these diverse *Archaea* have in common, what discovery led to identifying them as members of a coherent group, belonging to a single domain, one of only three great branches of the evolutionary tree: the *Eubacteria*, the *Eukarya*, and the *Archaea*. The most profound evidence for this tripartite tree, and especially for the place of the *Archaea* in it, is to be found in molecular structures, specifically in the sequences of nucleotide bases that make up DNA and RNA.

The First Step

On August 10, 1977, Carl Woese (pronounced "woes"), together with Ralph Wolfe and colleagues from the University of Illinois, submitted a short scientific paper to the prestigious *Proceedings of the National Academy of Sciences (PNAS)*. The paper, entitled "Classification of methanogenic bacteria by 16S ribosomal RNA characterization," was to appear in the October issue of the journal. This paper changed everything connected with the classification of microorganisms.

The singular importance of this paper was that it provided striking evidence that microorganisms could be classified, and their evolutionary relationships understood, by determining the structure of certain of their RNA molecules. This new technique classified organisms by comparing the sequences of nucleotide bases in RNA from their ribosomes (the subcellular particles in which proteins are manufactured). It turns out that two organisms exhibiting similar sequences are more closely related than two organisms with less similar ones, so that relatedness can be expressed in a quantitative fashion. The paper represented an early dividend of new techniques for determining nucleic acid sequences that were just being developed at that time.

Woese's breakthrough made it possible, for the first time, to construct a "tree" of evolutionary relatedness for bacteria, thus permitting a rational classification scheme. Moreover, the *PNAS* paper produced a second insight: organisms considered in 1977 to be bacteria actually belonged two distinct groups. One group contained the organisms called methanogens, the other, all the other "bacteria." The two groups were only distantly related: methanogens were no more closely related to the other "bacteria" than they were to, say, humans or oak trees.

The *PNAS* paper defined a turning point in evolutionary biology, both by illustrating the power of RNA sequences in defining evolutionary connections and shedding light on evolution of microorganisms. But the breakthrough had not occurred in complete isolation. For one thing, Woese and his colleagues had previously begun to employ RNA sequences as measures of bacterial relatedness, making considerable sense of what had been a rather murky topic. In fact, "murky" rather understates the situation that then existed in microbial classification: in 1977, bacterial classification had become quite unreliable, with major categories being redefined or disappearing with alarming regularity. Indeed, some microbiologists regarded bacteria as essentially unclassifiable and knowledge of bacterial evolution largely unobtainable. It seemed that the available criteria, such as cell shape, physiology, and metabolism, simply didn't work. They apparently didn't reflect evolutionary reality in any meaningful fashion.

There were already indications in the literature that methanogens exhibited unique features that set them apart from other "bacteria." For one thing, methanogens produce methane by means of biochemical pathways not found in any other sort of organism and, in doing so, employ some cofactors that are unique compounds not occurring anywhere else in the living world. In addition, methanogen cell walls lack compounds called peptidoglycans, which are almost universal components of bacterial cell walls. Intriguingly, certain halophiles were also observed to lack these peptidoglycans. Soon the halophiles would be linked to the methanogens on the basis of RNA analysis as well, so that the new category of *Archaea* would turn out to include both groups.

The Second Step

The *PNAS* paper was submitted by Woese, Wolfe, and their associates on August 10, 1977, and hit the newsstand, so to speak, in October. But a week after the first paper was submitted, a second paper—this time by Carl Woese and George Fox—was sent off to the same journal. This paper appeared in the November 1977 issue of *PNAS* and was entitled "Phylogenetic structure of the prokaryotic domain: The primary kingdoms." It was clearly an extension of the first paper and one really needs to think about them together. The message of the first communication was that RNA sequences can reveal microbial relationships, and, when they do, it becomes evident that methanogens are evolutionarily remote from other "bacteria." The methanogens thus constitute what the authors called a distinct domain.

The message of the second paper is much more global: based on RNA sequences, all organisms fall into three domains. These are: (1) the true bacteria or *Eubacteria*, (2) the *Archaea*, which Woese and Fox initially called Archaebacteria, and which include the methanogens as well as some others and (3) the *Eukarya*, which include all plants, animals, fungi, and many nonbacterial microorganisms.

A word in the title of the second paper, "prokaryotic," requires a preliminary definition: prokaryotic cells are small, simple cells that lack a membrane-enclosed nucleus. All bacterial cells are prokaryotic; until late 1977, the adjectives bacterial and prokaryotic could be used interchangeably. But it soon became clear that cells of the *Archaea* are also prokaryotic. In contrast, eukaryotic cells are structurally much more complex than prokaryotic cells, possessing the membrane-enclosed nucleus as well as other structures bounded by membranes.

A Matter of Style

Before 1977, no one had imagined that the living world could be classified in this tripartite fashion: Woese's RNA sequence approach had clearly led to unexpected and, as things turned out, enormously fruitful conclusions. It is interesting that this revolution was accomplished by the small group consisting of Woese, Wolfe, and a few associates. In contrast, most important scientific advances in this century have been the work of teams of scientists, often extremely large teams. But the discovery of the *Archaea* was a refreshingly intimate affair with Carl Woese and a few coworkers working in the context of the unsatisfactory state of bacterial classification, and, in the end, developing a new approach.

Admittedly, Woese had not invented the idea of using molecular sequences as measures of evolutionary relatedness. He was well acquainted with current literature describing the taxonomic use of amino acid sequences in proteins as well as nucleotide base sequences in DNA. But ribosomal RNA turned out to be an inspired choice, able to reveal relationships not evident with the other methods.

Unfortunately, not every inspired choice leads to immediate acceptance, let alone acclaim, by the scientific community: Woese was certain that his discovery of the tripartite classification of all life, and of the *Archaea* as one of the three great branches, was a major scientific achievement. He and his associates knew that they had changed biology in a fundamental way and thought that all biologists would immediately recognize and embrace the change. So one can appreciate their bemusement when, after an initial flurry of interest in the press, the advent of the "third form of life" was largely ignored by other scientists. They had proposed an alteration in the biological worldview that was evidently too major to be immediately acceptable. Everyone had been comfortable with a sort of dualism that considered all organisms to be either nucleus-lacking prokaryotes or nucleus-containing eukaryotes. A third form of life was definitely unwelcome.

From Woese's point of view, the worst thing of all was that no one rose to refute their work or deny its importance: there was just complete silence. This situation continued for a considerable period—the better part of a decade. But gradually the message began to sink in, and the voices of Woese and his few supporters came to be heard, so that now even introductory textbooks promote the triply branched tree of life. Textbooks have a way of being very conservative when it comes to new developments in science, the eventual inclusion of a new idea usually denotes widespread acceptance.

Thomas Kuhn in *The Structure of Scientific Revolutions* wrote that important changes in science occur through revolutions. These can be particularly painful because they amount to the overthrow of everyone's set of comfortable shared assumptions (called paradigms by Kuhn) and one psychological defense by other scientists is to pretend that the revolution never occurred. Eventually, this strategy necessarily fails and a new paradigm takes over. The discovery of the *Archaea* is perhaps analogous to Kuhn's revolution, with the three-domain organization of life constituting the new paradigm.

Ribosomal RNA Sequences

The data that led Woese and Fox in the direction of three domains hinged on the structure of ribosomal RNA (abbreviated rRNA). Organisms contain several classes of RNA, including messenger, transfer, and ribosomal RNA. All these RNAs consist of chains of alternating sugars and phosphate, with small molecules called nucleotide bases projecting from the sugars. There are commonly four different nucleotide bases: these constitute the alphabet with which the RNA "message" is written. The order of nucleotide bases, grouped in "words" of three bases, constitute the message, which, in fact, specifies the sequence of amino acids in proteins—each three-base "word" specifies the location of a particular amino acid in a protein. Of course, the central importance of RNA sequences in the discovery of the *Archaea* stems from the realization that sequence similarity between two organisms is a quantifiable measure of their degree of relatedness.

Why Sequences Differ in Different Organisms

It is well established that closely related organisms, having a recent common ancestor, have similar rRNA nucleotide sequences, just as they exhibit similar structures on other levels. Then, as the evolutionary clock ticks, structures and sequences diverge organisms come to resemble one another less and less. In the case of nucleic acids, the ticking of the clock represents random substitutions of individual bases in the overall sequence. These substitutions are also called mutations; because they are random, substitutions tend to be cumulative and they are known to occur at a fairly constant rate.

Actually, mutations are cumulative and constant in rate only if there is no environmental selection affecting them over time.

Thus, it is important that rRNA molecules are especially well insulated from environmental influences: their function is connected only with protein synthesis in general, not the synthesis of particular proteins. In other words, mutations in rRNA are neutral insofar as natural selection is concerned. For this reason, the rRNA mutational clock is assumed to be relatively unaffected by environmental vicissitudes and particularly constant in recording evolutionary time.

The Variety of Ribosomal RNAs

Ribosomes contain a number of different RNA molecules, varying in their size. This size is expressed operationally as the speed at which the molecule sediments in a centrifuge, using Svedberg units, abbreviated S. For instance, ribosomes contain 16S rRNA, with 1540 nucleotide bases, and 23S rRNA, with about twice as many. Because of their large sizes, the three-dimensional structure of rRNAs can be extremely complicated, often exhibiting a characteristic "cloverleaf" configuration.

Comparing Sequences

In 1977, methods for obtaining extended sequences of RNA molecules were just being developed, and Woese and Fox employed a relatively simple "fingerprint" technique. They used an enzyme called ribonuclease to cleave the RNA molecule into fragments, which they purified by electrophoresis, a technique that separates molecules on the basis of electrical charge. The fragments, called oligonucleotides, were much smaller than the intact 16S RNA, sometimes containing as few as six nucleotide bases. ("Oligo" denotes few.) Then Woese and Fox obtained base sequences for a collection of these fragments from specific organisms and termed the collections "dictionaries" for those organisms. Finally, they compared sequences of corresponding oligonucleotides in dictionaries obtained from a large number of organisms.

The comparison of sequences had to be strictly quantitative (and not just impressionistic). This was achieved by employing

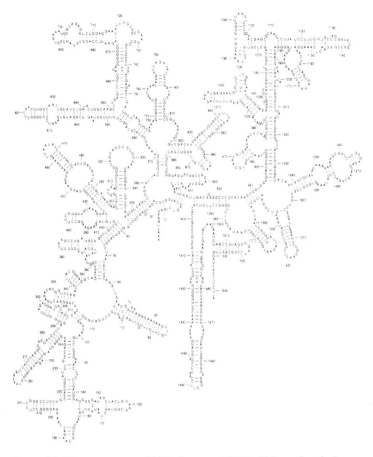

Figure 2.1 The structure of 16S ribosomal RNA. This molecule is from the eubacterium, *Escherichia coli*; corresponding RNAs from *Archaea* are very similar in overall structure.

an association constant, S_{AB}, as a measure of the similarity (the relative identity) of sequences from two organisms, A and B. This constant is defined as

$$S_{AB} = 2N_{AB}/(N_A + N_B)$$

where N_{AB} is the number of identical nucleotides in the pair of sequences and N_A and N_B are the total numbers of nucleotides in the sequences from organisms A and B. If all nucleotides are identical, then $S_{AB} = 1$. If the nucleotides are selected randomly from the four possible bases, then S_{AB} approaches a value of about 0.03. One carries out this procedure for all the oligonucleotides common to the dictionaries of two organisms and then obtains an average value for the S_{AB}. Therefore, it is possible to assign an exact number to any pair of organisms in the study and that number reflects evolutionary relatedness in a precise way.

Evolutionary Relationships from Sequence Data

In Woese and Fox's study, the average association constants, expressing the relationship between all the pairs of organisms included, were presented in a table with identical list of organisms on each axis—a table like the mileage tables in the margins of maps. It was evident that organisms fell into clusters of relatedness, as reflected by high association coefficients. Thus, in Table 2.1, one observes that the numerical values for S_{AB} fall into three distinct groupings. The organisms are arranged to emphasize the groupings, but the conclusions are not affected by that arrangement.

Look first at the cluster in the upper-left part of the table. Organisms *one* through *three* are the eukaryotic representatives, a yeast, a green plant, and a mouse. These exhibit high association coefficients with one another (about 0.3) and much lower values with all other organisms on the table. In a similar way, the block of S's in the center of the table indicates the high mutual relatedness among the *Eubacteria*. Notice that chloroplast rRNA appears to be eubacterial, one of many bits of evidence that chloroplasts, like mitochondria, evolved from bacteria by a process called endosymbiosis—engulfment of a microorganism by a host cell followed by evolution of the two in a mutually dependent state. And, finally, the block of high association coefficients in the lower-right portion of the table indicate the mutual relatedness of a cluster that turned out to be the *Archaea*, a group clearly demonstrating low affinity with the other two.

Thus, it was apparent that *Archaea* constituted a fundamental category of organism, the "third form of life." Actually, data for the *Archaea* in the original *PNAS* paper included only methanogens, but it soon appeared that some halophilic organisms

TABLE 2.1 Association Coefficients for Some Diverse Organisms

	1	2	3	4	5	6	7	8	9	10	11	12	13
1. *Saccharomyces* (yeast)	—	0.29	0.33	0.05	0.06	0.08	0.09	0.11	0.08	0.11	0.11	0.08	0.08
2. *Lemna* (plant)	0.29	—	0.36	0.10	0.05	0.06	0.10	0.09	0.11	0.10	0.10	0.13	0.07
3. L cell (animal)	0.33	0.36	—	0.06	0.06	0.07	0.07	0.09	0.06	0.10	0.10	0.09	0.07
4. *Escherichia*	0.05	0.10	0.06	—	0.24	0.25	.028	0.26	0.21	0.11	0.12	0.07	0.12
5. *Chlorobium*	0.06	0.05	0.06	0.24	—	0.22	0.22	0.20	0.19	0.06	0.07	0.06	0.09
6. *Bacillus*	00.8	0.06	0.07	0.25	0.22	—	0.34	0.26	0.20	0.11	0.13	0.06	0.12
7. *Corynebacterium*	0.09	0.10	0.07	0.28	0.22	0.34	—	0.23	0.21	0.12	0.12	0.09	0.10
8. *Aphanocapsa*	0.11	0.09	0.09	0.26	0.20	0.26	0.23	—	0.31	0.11	0.11	0.10	0.10
9. Chloroplast (*Lemna*)	0.08	0.11	0.06	0.21	0.19	0.20	0.21	0.31	—	0.14	0.12	0.10	0.12
10. *Methanobacterium sp. 1*	0.11	0.10	0.10	0.11	0.06	0.11	0.12	0.11	0.14	—	0.51	0.25	0.30
11. *Methanobacterium sp. 2*	0.11	0.10	0.10	0.12	0.07	0.13	0.12	0.11	0.12	0.51	—	0.25	0.24
12. *Methanobacterium sp. 3*	0.08	0.13	0.09	0.07	0.06	0.06	0.09	0.10	0.10	0.25	0.25	—	0.32
13. *Methanocarcina*	0.08	0.07	0.07	0.12	0.09	0.12	0.10	0.10	0.12	0.30	0.24	0.32	—

Source: Adapted from Woese and Fox (listed in Additional Reading).

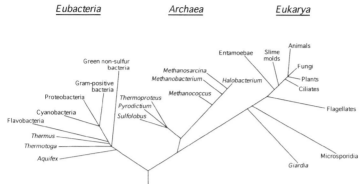

Figure 2.2 The phylogenetic tree of all living things. The tree is based on rRNA sequence comparisons in such a way that evolutionary distance between two organisms is proportional to the lengths of the lines that connect them. This tree, as well as the one in Figure 2.4, is adapted from those of Woese. (See, for example, his introductory chapter in Kates et al., listed in Additional Reading.)

(such as *Halobacterium*) and thermophilic organisms (such as *Sulfolobus* and *Thermoplasma* were in the same cluster of association coefficients. Thus, the *Archaea* came to include all three "lifestyle" groups, methanogens, halophiles, and thermophiles. So it was that this rRNA sequence-based cluster of microorganisms, the *Archaea*, began to take shape, bringing together organisms that were formerly believed to be evolutionary unrelated and only linked by their distinctly odd lifestyles and stringent environments. Such stringent niches would be expected to foster evolutionary conservatism: creatures living there would evolve relatively slowly so that *Archaea* were apparently both very tough and of ancient stock. Indeed, we now know from rRNA studies that the *Archaea* are the most slowly evolving organisms on earth.

The Felicitous Choice of 16S rRNA

That particular 16S RNA molecule was a good choice for several reasons. For one thing, it is found in all organisms, and it always has the same function—participation in protein synthesis. We saw that this role effectively removes the molecule from the more immediate effects of environmental selection that could obscure long-term, cumulative changes. For another, the 16S molecule turned out to be of a particularly appropriate size for the studies in question. It is a sufficiently large molecule to carry adequate information for detailed evolutionary studies. Smaller 5S rRNAs are simply too small to be useful, and their sequences can also be too unstable. On the other hand, 23s rRNA was unmanageably large for the early studies although, with current technology, it can now be actively exploited.

A particularly elegant feature of 16S rRNA turned out to be the manner in which different parts of the molecule exhibit different rates of evolution, as measured by base substitution. Thus, some parts of the molecule undergo very infrequent base substitutions and so can be used to measure large evolutionary distances—long time intervals in the history of life. Other regions of the molecule change more rapidly (in evolutionary time) and are useful in studying shorter time spans. This feature of the molecule has been likened to a watch with both an hour and a minute hand.

The Idea of Domains

On the basis of cell structure, there are two fundamental sorts of organism, prokaryotic and eukaryotic. But, based on rRNA sequences, there appear to be three fundamental sorts, the *Eubacteria*, the *Archaea*, and the *Eukarya*. This has led to something of a dilemma in classification. An approach to the problem is Woese's introduction of a new category, the domain: the most inclusive of phylogenetic groupings, even superseding the familiar kingdoms. Accordingly, all organisms belong to one of three domains, *Archaea, Eubacteria*, and *Eukarya*, the first two including prokaryotic cells, the third, eukaryotic. And within the domains are kingdoms. For example, Woese has proposed that the *Archaea* include two kingdoms (described below); the *Eubacteria*, several; and the *Eukarya*, at least four; the list is probably incomplete. Woese and his colleagues predicted that the number of recognized kingdoms will tend to increase over time.

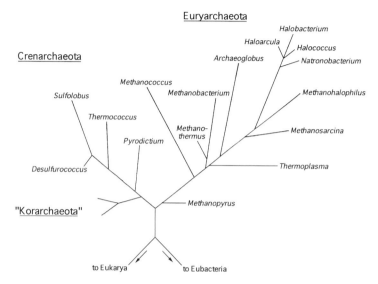

Figure 2.3 The phylogenetic tree of the *Archaea*.

More rRNA Studies: Relationships within the Archaea

Having established that all life can be partitioned into three (and only three) distinct domains, rRNA research has turned to evolutionary groupings *within* the *Archaea*. Based on lifestyle, one speaks of three kinds of *Archaea*: methanogens, halophiles, and thermophiles. However, looking at rRNA sequences, these are definitely not the groupings that emerge. To understand this, it is helpful to depict sequence information in the form of a tree or dendrogram, a kind of graph that represents per-cent differences in RNA sequence between two organisms as distance.

Figure 2.3 shows such a tree for the *Archaea*: it is immediately clear that there are two, not three, main branches. One of the two branches contains only the thermophilic, sulfur-dependent *Archaea*. For the moment, let's call this branch Group I. The other, Group II, includes everything else: all the methanogens, all the halophiles, and even a few thermophilic organisms. This group also includes organisms that are both thermophilic and methanogenic, so that the ability to thrive under hot conditions is not restricted to Group I. Furthermore, Group II consists of three main clusters. Two of these contain methanogens only, whereas the third is a mix of methanogens, halophiles, and a few thermophilic sulfate reducers. This pattern of relationships was initially derived on the basis of 16S rRNA, but was later confirmed by studies with 23S rRNA. Finally, there are a few additional organisms, such as *Methanopyrus*, that branch deeply from the others and so don't fit very well in either group.

Archaeal Names

The *Archaea* are still so new that their nomenclature has not been solidified by long usage. Even *Archaea* is not universally employed; some biologists still call them by their original name, "Archaebacteria." Of course, this earlier name suffers from the defect of implying that they are, after all, just some type of bacteria—which they emphatically are not. Both names are based on the same Greek root that gives us "archaic" and "archaeology," words that smack of the ancient. But, in fact, the earliest Greek meaning is probably slightly different: "to begin." Thus, one imagines that Carl Woese sought to convey "the beginning of life" when he coined Archaebacteria to denote this new group of microorganisms. It turned out to be an astute choice, given the slow pace of archaeal evolution.

Archaeal "Group I" is now called the *Crenarchaeota* and "Group II," the *Euryarchaeota*, the two kingdoms that Woese proposes as constituting the *Archaea*. Not everyone accepts these kingdoms and there are some alternative taxonomic terms floating about. "Eocyte" is one example; it reflects a different interpretation of the sequence data and may be almost, but not quite, synonymous with *Crenarchaeota*. In fact, the two kingdoms of Woese may only be the beginning. Norman Pace and associates have found rRNA sequences in Obsidian Pool, an anaerobic hot spring in Yellowstone National Park, that do not appear to belong to either the *Crenarchaeota* or the *Euryarchaeota*. These sequences appear to form the earliest known branch of the archaeal tree and their discovers propose a new kingdom to accommodate them, naming it *Korarchaeota* for the Greek, *kore*, young woman. Unfortunately, this new kingdom is manifest chiefly as nucleotide sequences, and no living organisms belonging to it have yet been isolated.

The Archaeal Tree

Here are some general observations about the archaeal tree. First, all evidence points to a thermophilic origin for the *Archaea*. The thermophilic way of life is widely distributed in all the branches of the *Archaea*, including even the deeply branched (i.e., ancient) *Methanopyrus*. Thus, it is likely that *Archaea* appeared at a time in the history of the Earth when there were widespread hot (geothermal) locations available. Similar evidence indicates that *Eubacteria* also first evolved in hot places and, significantly, that *Eukarya* didn't.

It also seems that at least the halophiles and sulfate-reducers, and perhaps also the thermophiles of the kingdom, *Crenarchaeota*, arose from a methanogenic lineage. Here, the place of *Methanopyrus* seems pivotal: this deep-sea archaeon is located very near the root of the archaeal tree, is extremely thermophilic, growing at a temperature up to 110 degrees, and is a methanogen, making methane from carbon dioxide and hydrogen. Therefore, it is likely that the first *Archaea* were methanogenic and the ability to make methane may be the primitive form of metabolism in the *Archaea*. Or, at the very least, the split between methanogens and sulfur reducers may be close to the archaeal root—in which case, methanogenesis would not necessarily be primordial, but merely quite ancient. Unfortunately, we really don't yet have enough evidence to be sure of this point. Among other things, we could really use more information about the proposed *Korarchaeota*: for example, are they methanogenic?

Getting at the Root of It All

There is good reason to believe that all living organisms have descended from a single, common ancestor. (Convincing evidence for this assertion includes the universality of the genetic code, ribosomes, and lipid membranes.) Thus, one is entitled to wonder which living organisms are most directly descended from that last common ancestor. This amounts to locating the root of the universal evolutionary tree.

Note that the last common ancestor was itself the product of relatively long evolution, stretching back to the very first life. That common ancestor probably exhibited most of the important attributes of a modern prokaryote, and would have offered few clues about the characteristics of its really primitive precursors. Indeed, Carl Woese has recently speculated that the common ancestor may have evolved not from a discrete precursor at all, but from a population of "protocells" with simple genetic systems and very limited metabolic capabilities. These primitive biological entities would only be viable as a community and would require one another, for example, to carry out complete metabolic processes. They would readily exchange genetic material, owing to their relatively unsophisticated genetic systems. For this reason, the community would evolve as a unit and so could serve, in subsequent evolution, as a precursor of proper organisms, leading finally to the last common ancestor. Moreover, the first life probably produced other evolutionary branches that were "frustrated" by cataclysms like meteor impacts or widespread climate perturbations, thus leaving no line of descendants. But such "unproductive" evolutionary lines constitute a closed book for us. So, from our vantage point, we can only see really as far as the last common ancestor. All else is conjecture.

Unfortunately, there is a problem associated with rooting a universal tree. This tree is based entirely on comparisons of rRNA base sequence between pairs of organisms included in the tree, with distance along the branches reflecting the number of base substitutions when comparing the two organisms. However, at the root of the tree, there is no other organism with which to compare: one needs an "outlier"—a comparison group—and there simply is none. This dilemma has been solved in a most ingenious way. Certain genes occur as multiple copies in an organism, and these copies can have an evolutionary life of their own, evolving independently of one another. Fortunately, the duplication of some of these genes occurred at a very early time in evolutionary history. The independence of the multiple gene versions makes it possible to obtain values for distance between two, (or more) copies in the same organism. Thus, one can, in effect, use an organism as its own outlier and, in that fashion, obtain a root.

Using this approach, it becomes possible to see how the three great domains, the *Eubacteria*, the *Archaea*, and the *Eukarya*, connect to one another and to their common root (Figure 2.2). We discover that the two prokaryotic domains are about equally close to the root, whereas the eukaryotic organisms diverged much later. Clearly, both prokaryotic groups are of great antiquity and, as one might expect, diverged from one another a very long time ago. However, the approach led to an additional result that no one expected: rRNA analysis showed the *Eukarya* to be much more closely associated with the *Archaea* than the *Eubacteria*. This association has serious implications for the evolutionary history of the *Eukarya*; we will consider these implications in Chapter 8.

The Problem of Convergent Evolution

Evolution proceeds largely, although not exclusively, by means of natural selection. That is, adaptation to the environment, and especially, to environmental change, is a major evolutionary driving force, and the genes of organisms that adapt successfully are progressively selected. However, quite distantly related organisms can sometimes adapt in quite similar ways, independently "discovering" the same evolutionary solution to the same problem. This can lead to similar structures in quite distantly related organisms. The human and the octopus eye have many features in common, although the two organisms are evolutionarily remote. In such cases, we say that remote lineages *converge* on the same solution to an environmental requirement, leading to the term "convergent evolution." In this example, similarity in eye structure might lead to a entirely erroneous decision as to the classification of the two organisms in question; presumably, other structural features would be more useful, but how does one decide which of these to select?

Convergent evolution becomes a particular problem when physiological and biochemical traits are employed as indicators of evolutionary relationships. The difficulty is knowing what sort of feature to select—making sure that attributes to be employed truly reflect evolutionary distance only, and not convergent adaptation to the accidents of the environment. This difficulty is especially noticeable when one tries to use nutritional features of microorganisms to place them in an evolutionary scheme. Microorganisms are simply too ad hoc when it comes to nutrition and their adaptability obscures evolutionary patterns.

Classification of Microorganisms

For many years, efforts to classify prokaryotes led, as we saw, to rather unsatisfactory results, as evident in the wild taxonomic swings exhibited by consecutive editions of *Bergey's Manual of Systematic Bacteriology*, the authoritative guide to such matters. For one thing, classification was heavily based on nutrition and associated metabolism, an approach that led to organisms that we now understand are closely related being classified as remote, and vice versa.

Physical structure had always been the mainstay of efforts to classify the so-called higher organisms. For instance, the skeletal structure of vertebrate animals has usually provided excellent guidance in defining relationships. But among the prokaryotes, there is simply too little observable structure and too much similarity among different organisms in whatever structure can be observed. Bacteria can be separated, for example, into cocci (spherical cells) and bacilli (rod-shaped cells), but such distinctions have little to do with evolution and indeed the shape of cells often changes dramatically with alterations of the growth conditions of the moment. The structure of microorganisms is simply too limited and much too variable to provide reliable guidance evolutionary relationships.

In addition, prokaryotic organisms seldom leave unambiguous fossils. Large eukaryotic organisms may be localized in evolutionary history even after they have become extinct; their imprint often persists in the form of fossils that can be highly informative. Such fossils may often be dated by a radioactive decay "clock," or by knowledge of their geological surroundings, so that it is possible to specify their occurrence in a strict chronology. Unfortunately, prokaryotic fossils are microscopic, notably lacking in detail, and often difficult to identify. Indeed,

it is sometimes difficult to be sure that they are fossils at all—
that they have ever been alive—and often they are interesting
only in the sense that they occur at all. For example, some
"microfossils" have been found in rocks dated at about 3.5
billion years ago, thus establishing a time when prokaryotic life
was certain to have existed. However, one can't tell much about
the organisms in question, except that they were extremely
small and, in some vague respects, resembled modern bacteria.
Likewise, fossil bacterial colonies called stromatolites (fossilized
microbial mats) are of similar great antiquity and suggest that
symbiotic associations between different organisms have been
around for a long time. But, again, they disclose little structural
detail to gladden the heart of a student of evolution.

Molecular Classification Prior to rRNA

For many years, there had been interest in using molecular
structures in bacterial classification. An early manifestation of
this focus was the effort to use the nucleotide composition of
DNA as a taxonomic tool. The composition was commonly expressed as the guanine plus cytosine (G + C) ratio of a particular organism, the "GC ratio." This is the fraction of base pairs
that are GC, in contrast to A + T (adenine + thymine). It is
expressed as a percent:

$$\frac{G + C}{G + C + A + T} \times 100 \text{ percent}$$

In various microorganisms, the value ranges from about 25
to 65 percent and, in some instances, can be used to characterize particular organisms. For example, the bacterium *Desulfurococcus* closely resembles a number of other terrestrial thermophiles in structure and physiology, but can be readily
separated from them by its relatively high GC ratio, 51 percent,
as compared to values generally in the 30 to 40 percent range.
Indeed, when new microorganisms are described, the GC ratio
is a required part of the description, along with structural and
nutritional characteristics. However, the ratio does not necessarily reflect base sequence and did not turn out to be very
helpful in most aspects of classification—that is, in uncovering
evolutionary relationships.

Fortunately, at about the same time, protein structure offered greater promise. Thus, it became evident in the 1960s that the sequence of amino acids in proteins could, in principle, be used as a measure of relatedness. Such amino acid sequences are clearly defined, contain large amounts of information, and support great precision and consequent ease of mathematical analysis. In most proteins, there are about twenty different amino acids, and a single protein can contain hundreds of them. The amino acids of a specific protein are arranged in a unique sequence, which determines the protein's overall (three-dimensional) structure and affects every aspect of its function.

The argument should be familiar by now: comparable proteins in different organisms exhibit at least some similarity in the sequence of their amino acids. The more closely the organism are related, the more similar their sequences often prove to be. For example, a particular enzyme might exhibit an identical sequence in humans and gorillas, with only a few amino acid substitutions between them and some lower primate. Additional substitutions would be observed as one compared enzyme sequences from more evolutionary distant sources: perhaps the majority of amino acids would differ when comparing, say, the human/gorilla protein to that from some lowly bacterium.

Unfortunately, amino acid sequence analysis turned out to be rather disappointing in the case of bacteria. The great adaptability of bacteria led to their proteins being too variable, as the vagaries of the environment exerted too great an influence on protein structures. Evolutionary trends were sometimes apparent, but these were usually confounded by so much unhelpful "noise" that detailed conclusions were often quite impossible.

If amino acid sequences can reflect evolutionary relationships, so must the base sequences of the DNA that encoded them. The use of DNA in studies of relatedness actually predated the ability to determine explicit nucleotide base sequences, with mutual binding of DNA strands being used as a measure of similarity in sequence. This is possible because the two strands of a double-stranded DNA molecule are *complementary* to one another—the bases form cross-strand pairs, based on hydrogen bonding, and such complementary binding physically holds the two strands together in a double helix. The degree of complementary determines the tightness of binding between the two strands. Therefore, closely related organisms, with their closely similar sequences, exhibit tight binding, which can be used as a measure of their relatedness. In practice, DNA strands are separated by careful heating after which reunification is measured at a lower temperature. One follows the process of separation and restoration of the double helix by monitoring changes in ultraviolet light absorption, which differs between the single-stranded and double-stranded states. This method was applied before 1977 to a variety of microorganisms; it revealed unexpected similarities as well as lack of relatedness, providing a preliminary indication that revision of microbial classification would soon be required.

Classification Problems Resulting from Gene Transfer

Initial use of either protein sequences or DNA binding in microbial classification was encouraging, but not completely successful: evolutionary trends were often apparent, but with intolerable variability. It turned out that much of the variability originated in the unexpected ease with which many microorganisms transfer genetic information (in the form of DNA), a feature that accounts for their great adaptability. There are several mechanisms for the transfer of information-bearing DNA among bacteria of the same, or different, species. These include transformation, which is cell-to-cell transfer of naked DNA fragments, and transduction, which is the transfer of bacterial DNA packaged in virus particles. Organisms can also share their plasmids, which are small, circular DNA molecules containing, at most, several genes, and many carry out conjugation, which is the donation of all, or part, of a chromosome by one cell to another. Such transfer occur between even quite unrelated bacteria, even between *Eubacteria* and *Archaea*. Such "lateral" (or "horizontal") gene transfer can introduce novel gene sequences in bacteria in a fashion that has nothing to do with evolutionary relatedness, and it can make resulting protein sequence data extremely difficult to analyze.

For example, the amino acid sequence of cytochrome c, a small, easily purified protein, was studied in a wide selection of organisms, including a number of bacteria. This work tended to confirm what was already known about relationships between eukaryotic organisms but didn't do much for bacterial classification, owing, presumably in part, to the effects of lateral DNA transfer. Such protein studies indicated that bacteria were much more variable than eukaryotic organisms: among the bacteria, a genus exhibited more variability in cytochrome c structure than did an entire class, a much larger taxonomic category, of "higher" organisms. It was as though the deck was constantly being shuffled while one tried to make predictions about the cards. And similar factors worked against the usefulness of DNA binding in studying bacterial evolution, so that the evolution and classification of microorganisms remained, for many years, largely a closed book. This was, of course, a great pity, since, for much of the 4.8 billion years of evolutionary history, microbial evolution was all that there was.

In fact, this unsatisfactory state of bacterial classification led, almost three decades ago, to a search for more reliable criteria for relatedness, that is, a more accurate clock to measure the pace of evolution. Such a timepiece would have to be relatively isolated from environmental vicissitudes, changing only according to its own internal schedule. Fortunately, rRNA sequences provide chronometers of this sort and their use has begun to introduce excellent sense into bacterial classification.

RNA sequence studies also produced a number of surprises, such as the discovery that photosynthetic bacteria did not constitute a coherent group. Instead, some of them turned out to be more closely related to *Escherichia coli*, a highly adaptable bacterium that often lives in animal intestines, than they were to other photosynthetic creatures. But we saw that the biggest surprise of all occurred when Woese and associates began to use the newly developed RNA sequence approach to examine methanogenic "bacteria."

Ribosomal RNA is insulated from environmental influences so that rRNA sequences evolve very slowly and at a constant rate, with their base changes reflecting only random substitutions. Moreover, the suitability of rRNA as an evolutionary chronometer is enhanced by being protected from lateral DNA transfer—for example, via transformation or the sharing of plasmids. This is because the genes that encode rRNA sequences occur as multiple copies in the cell's DNA. Their multiplicity tends to dilute corresponding DNA transferred from external sources. This redundancy is presumably a reflection of the great importance attached to protein synthesis and the dire consequences resulting from any perturbation of it.

Using rRNAs to Inventory Microorganisms in the Wild

Microbial ecologists have traditionally characterized natural populations by obtaining samples from the environment and trying to culture the resident organisms under controlled conditions. This strategy inventories only those organisms that can be made to grow under the particular conditions employed and it has recently become apparent that it samples only a small fraction of the organisms that actually live in environments such as soils or oceans. Indeed, it seems that only a few percent of the total number of resident species can be grown under laboratory conditions. The others simply don't grow under the conditions that we know how to provide, or are swamped by competing organisms that are favored by laboratory conditions. In some instances, microorganisms live in obligatory symbiotic associations of such strict interdependency that they cannot be isolated singly.

In fact, studies of rRNA sequences have uncovered this large fraction of microorganisms that resist being grown in the laboratory: it has proved possible to identify, in nature, large numbers of rRNA sequences that do not belong to any previously cultured organisms. Such unknown sequences can often be identified as belonging to a certain class of microbe, such as photosynthetic *Eubacteria* or one of the various categories of *Archaea*. So these sequences take on a life of their own and serve as a sort of proxy for organisms that may be difficult (or impossible) to study, as such. It is therefore possible, in principle, to contemplate ecological studies that are quite "uncontaminated" by any living organisms and which, in fact, use only molecular data to represent the microorganisms resident in a real ecosystem.

Here is one dramatic example of the use of RNA sequence analysis in an environmental situation. A thermophilic archaeon was isolated from a Yellowstone hot spring after its existence there had earlier been predicted from 16S rRNA data.

In this study, a mixture of organisms was obtained from hot-spring water and grown in the laboratory. A fluorescent stain was prepared that contained an RNA sequence complementary to an archaeal rRNA base sequence belonging to the hitherto unknown organism in question. The RNA from the unknown beast had earlier been obtained by chemically extracting all the nucleic acids from the environmental sample, followed by chemical amplification and analysis of sequence (employing techniques to be described shortly). In this manner, it was possible to obtain sequences without ever having isolated the actual organisms that contained them. The stain, containing the complementary sequence, would only bind to cells of the species that contained the original sequence. It turned out that an extremely minor member of the population, an organism that grew in grape-like clusters, exhibited fluorescence, and it was possible to isolate it as a pure culture. Thus, the unknown organism, initially corresponding only to a base sequence obtained from nature, materialized as a real, living entity.

The fluorescent stain approach has also been used to visualize living cells of the hypothetical *Korartchaeota*, until now only known through nucleotide sequences. Fluorescent stains incorporating those sequences were added to a mixture of organisms from the hot spring in Yellowstone National Park, where the sequences were initially obtained. A small fraction of the microorganisms bound the stain, identifying them as *Korarchaeota*. It has not yet proved possible to grow them, free from other cells, under laboratory conditions, so we know little about them.

A final example of the use of sequence-specific fluorescence staining in the search for new organisms is the discovery of thermophilic-like (i.e., *Crenarchaeota) Archaea* inhabiting a marine sponge. Although identified by its rRNA sequences as a member of the thermophilic category, the archaeon grows well at 10 degrees indicating a more varied membership in the *Crenarchaeota* that had been previously thought. And although, in some other cases, there appear to be variable consortia of microorganisms inhabiting a particular marine invertebrate, in this instance, there is the unique association between a particular archaeon, *Cenarchaeum symbiosum*, and a particular sponge, *Axinella*.

Environmental DNA Used to Identify Organisms

The technique of using environmental rRNA sequences requires amplification of the small amounts of nucleic acid from the organisms found in nature. In most instances it is not rRNA itself, but rather the DNA encoding the RNA sequence, that is amplified. This is possible because the base sequence of a molecule of rRNA was initially encoded in a gene composed of "rDNA." Obviously, "ribosomal DNA" is something of a misnomer since it is not located in the ribosome. Because rRNA is synthesized using the rDNA as a template, there is an explicit (complementary) relationship between sequences in rRNA and in the corresponding rRNA, so that either can be used to identify an organism.

Here is how the identification works. Samples of soil, water, or other material are obtained from the environment and the total complement of DNA present is extracted, using a suitable solvent. This extract contains a mixture of the DNA from all the organisms present in the sample. Care must be taken that the DNA molecules are not degraded because the procedure requires genomic sequences that are as intact as possible.

The rDNA is then chemically amplified to yield numerous copies of each original molecule. This is accomplished by the "polymerase chain reaction" that has acquired so much importance in molecular biology that it is invariably abbreviated PCR. The PCR procedure imitates the living cell's way of replicating nucleic acids, using natural enzymes to catalyze the process. The procedure also employs an automated program of heating and cooling to separate and reform double-stranded DNA; the program is carried out by a device called a thermal cycler.

The steps in the PCR process are as follows. The procedure begins with heating to separate the DNA into single strands. This is termed the denaturation phase of the cycle and typically (for rRNA genes) requires ninety seconds at 92 degrees. Then, the solution containing single-stranded DNA is cooled to 50 degrees for ninety seconds in the presence of chemically synthesized DNA "primers." This is called the annealing phase of the cycle; annealing means that hybrids of DNA and primer are formed by base pairing. Each of the pair of separated DNA strands binds its own primer; the locations of these two primers form the limits of the DNA to be amplified. These primers are small, obviously much shorter than the single-stranded DNA to which they are bound. It is clear that they must be complementary to a small region of the DNA for base pairing to occur. In principle, one must know a portion of the DNA sequence to synthesize a suitable primer in the first place. A requirement for successful PCR amplification is the selection of suitable primers. Primers range in size from about fifteen to eighty nucleotide base pairs, with longer primers requiring a higher annealing temperature and being considerably more specific as to organism and sequence being amplified.

Next, the enzyme, DNA polymerase, is used to catalyze the extension of the primers, again according to the rules of base pairing. In the case of 16S rRNA, this requires two minutes at 72 degrees, a temperature intermediate between those used for denaturation and annealing. The extension of strands complementary to the two original DNA strands occurs in opposite directions, in both cases, starting from the respective primers. At this point, there are two complementary strands forming a double helix and the cycle can be repeated to obtain the desired degree of amplification. This overall process can entail thirty cycles of replication, requiring several hours and producing a yield of as much as 10^9 fold amplification of the original sequence. The primers that play such a central role are often specific for particular groups of organisms, or even for particular genes, so that they can be used to make amplification much more specific. For instance, by selecting the right primer, one can copy mostly rRNA genes, and not any of the huge numbers of other genes in the organism (or collection of organisms). Indeed, one can specialize further, and copy primarily the rRNA genes from *Archaea* and not, for example, from eukaryotic organisms. Conversely, if one seeks to amplify an unknown sequence, as from an unidentified organism, universal primers can be employed. These, containing a consensus of many sequences, work with a wide phylogenetic range of organisms.

Finally, DNA polymerase deserves additional mention. The polymerase is a naturally occurring enzyme, isolated from a suitable microorganism. In this instance, "suitable microorganism" has a special meaning: in the PCR cycle, the polymerase must endure the high temperature required to separate the DNA strands. However, enzymes are usually heat-sensitive and the polymerase would normally be in danger of destruction at elevated temperatures. So it turns out that DNA polymerase from the thermophilic eubacterium *Thermus aquaticus* is just the thing. This bacterium, which we have already encountered as one of the first thermophiles studied by the Brocks in the 1960s, grows well at a temperature close to that of boiling water, and its enzymes survive and function under such conditions. Happily, its DNA polymerase, called Taq-polymerase, is stable up to temperatures of 95 degrees and so can easily tolerate the rigors of the amplification process.

When significant amounts of rDNA have become available though PCR, it then becomes possible to determine the sequences of particular rDNAs using routine sequencing methods. In this way, one can obtain very large collections of such sequences. Each distinct rDNA sequence corresponds to a distinct organism and, if one is fortunate, one can identify them through reference to known sequences.

Determining DNA Sequences

DNA is routinely sequenced in an automated fashion, using a method exhibiting certain similarities to the PCR technique. Thus, both methods begin with single-stranded DNA, both require suitable primers, and both employ DNA polymerase to make double-stranded DNA. The most widely used approach to sequencing is called the chain termination procedure, with "termination" referring to termination of DNA replication by artificial analogs of normal DNA synthesis ingredients. Normally, DNA is synthesized using four deoxyribonucleoside triphosphates (abbreviated dNTP), each consisting of a nucleotide base, such as adenine of thymine, as well as the sugar deoxyribose and a chain of three phosphates. A mixture of these four dNTPs, single-stranded DNA, a primer, and DNA polymerase will yield double-stranded DNA as large as the single-stranded DNA template—that is, a second strand will have been synthesized along the first. However, analogs of dNTPs, in which the sugar is lacking an oxygen—dideoxynucleoside triphosphates—do not support the synthesis. When one of them is added to the mixture, the synthesis terminates because subsequent nucleosides cannot add to the dideoxy one. Therefore, the dideoxy analog is always the last base in the terminated chain.

The sequencing procedure begins with a reaction mixture containing the single-stranded DNA to be sequenced, the primer, all four dNTPs, one of which is radioactively labeled and one of the dideoxy analogs. More exactly, the experiment is carried out in four separate reaction mixtures, each including a different analog. In each separate reaction mixture, DNA is replicated by the polymerase until terminated by one of the analogs, producing DNA pieces that end with that analog. The pieces from the four reaction mixtures are then separated by a type of gel electrophoresis, in which small pieces move further in the gel than large ones.

Each reaction mixture contains a mixture of pieces of different sizes; each ends with the analog used in that particular mixture and that the smallest possible piece is one base long, the next smallest, two bases long, and so on. If, on examining the gel, one observes that the smallest piece is from the reaction mixture that contained dideoxyATP, the second smallest from one that contained dideoxyTTP, and the third smallest from one that contained dideoxyGTP, one has discovered that the order of bases is adenine–thymine–guanine. Thus, one reads the order of bases directly from the gel, a process that can easily be automated, and it becomes possible to determine the sequences of extremely large spans of DNA. This chain-termination sequencing technique underlies a great deal of molecular biology and permits not only sequence evaluation of relatedness, but also the massive genome projects that are so enlivening contemporary biology.

Two other points require mention. First, the location of pieces of newly synthesized DNA on the gel is disclosed by autoradiography: the radioactivity from the labeled base produces spots on a firm placed against the gel and subsequently developed. Finally, the order of bases determined by this sequencing procedure is actually the order of bases in the new strand that has been synthesized, so that one must use the rules of complementarity (base-pairing) to know the order on the original single-stranded DNA.

Archaea Are Everywhere

From such studies, it has become evident that there are large numbers of unknown organisms in any environment and that only a very small fraction of DNA sequences correspond to known varieties. It is also apparent that the *Archaea* are much more widely distributed in the biosphere than had been suspected. For example, sequences similar to those from highly thermophilic *Archaea* occur widely in cold seawater, in samples of ordinary soil, and, as we saw, in marine sponges. Science has only begun to appreciate the variety and wide distribution of the *Archaea*; archaeal biology is clearly in its infancy.

Indeed, Microbes Are Everywhere

Prokaryotic microorganisms of all sorts are also even more ubiquitous and more diverse than previously thought. Formerly, estimates of microbial populations, and judgments about their ecological significance, were based chiefly on culturing samples from the environment; such studies revealed large and diverse assemblages in virtually all kinds of habitats. However, culturing missed organisms that didn't grow under the exact conditions tried, and one also tended to underestimate the variety of microbes owing to an inability to tell many of them apart. But advances in rRNA analysis put an end to those difficulties, both by providing an unambiguous way to specify an organism and making possible the provisional identification of the sizable percent of organisms that resist being grown in culture. Thus, we were previously unaware that prokaryotic microorganisms (including *Archaea*) constitute a major fraction of the planktonic communities in the ocean and that *Archaea*, at least, also live in large populations in deep fissures in the Earth's crust. Microorganisms—archaeal and eubacterial—certainly reside in more varied habitats than any other organisms: their environments range from hot marine vents to the human intestine. Perhaps because we ourselves inhabit so restricted a selection of the Earth's habitats, we tend to underestimate the full range of locations where microorganisms reside.

Microorganisms also constitute a significant fraction of the world biomass, perhaps to the extent of exceeding all other organisms and, without question, they play a central role in the chemical transformations of the world ecosystem. Indeed, there is also considerable evidence that microbial populations are responsible for creating, or transforming, many of the minerals, soils, and gases that make up the planet's crust and render it habitable for other forms of life. Some "banded" iron ore formations clearly show their microbial origins, and bacteria are even thought responsible for the precipitation of gold and other metal nodules from seawater. It is only because they are individually so small that we largely manage to ignore microorganisms. Stephen Jay Gould, whose name is usually associated with the study of somewhat larger organisms, has said that we live in an "age of bacteria" and, indeed, that it has always been so: bacteria—and *Archaea*—have dominated the earth through most of its history and for the first two billion years of life, they had a monopoly.

Voyages of Exploration

People interested in natural history often complain that the "age of exploration" is over. It might appear that the biosphere is well known and thoroughly picked over. It may in fact be true, in this era of satellite mapping and global travel, that some of the fun has gone out of old-style exploration, but readers of this chapter must realize that it is too early to despair. Especially where microorganisms are concerned, wonders remain undiscovered.

Finding Them in Nature
(and Bringing Them Home)

Getting to know microorganisms well requires studying them under controlled conditions, generally in a laboratory setting. Therefore, they must be plucked from their usual habitat (nature) and somehow induced to grow in culture, preferably without competing species. This procedure, called isolation, can require considerable strategy—for example, in developing isolation media based on a reasonable judgments about the metabolic capabilities of target organisms. (In this fashion, isolation procedures link metabolic biochemistry and ecology.) Indeed, it sometimes appears quite difficult; recall that some microorganisms known to exist in nature have resisted all attempts at isolation and, as a consequence, we don't know very much about them. And when we (somehow) do learn more, for example, about their metabolism, then it often becomes much easier to obtain them in culture.

We observed that discovery of the archaeal domain was based on a comparison of nucleotide sequences, using techniques like RNA and DNA isolation, polymerase chain reaction (PCR), and chain-termination sequencing. In addition to nucleic acid sequences, other attributes that set the Archaea apart from other organisms have been chiefly biochemical, such as the presence (or absence) of certain cell-wall constituents, membrane components, or exotic coenzymes. Such matters may appear somewhat arcane and remote from any possible "real life" of these microorganisms. The Archaea obviously do have a real life in nature, but to really get to know them, one must bring them home and study them under the controlled conditions of a laboratory.

The World of Microorganisms

We humans live in a world surrounded by microorganisms, but, for the most part, are blissfully unaware of their presence and mostly ignorant of their massive role in the biosphere. "Microorganism" is a catchall word for any creature too small to see with the unaided eye; microorganisms include all *Archaea* and *Eubacteria*, as well as many members of *Eukarya* (e.g., protozoans and yeasts). Microorganisms live in virtually all of the Earth's habitats and dominate many of them. And microorganisms chemically transform their surroundings in innumerable ways, profoundly affecting other organisms and acting as significant geochemical forces.

Microbes form a major fraction of the mass of any soil sample; the fertility of the soil is profoundly influenced by their presence. Likewise, microbes inhabit all bodies of water with varying population densities. (When densities are high, we often view the microbes as symbols and criteria of pollution, forgetting that they are also important agents of remediation.) Finally, entire ecosystems of microorganisms, chiefly bacteria, live in and on our bodies; we are mostly unaware of these companions unless the ecological balance goes awry. In other words, we share our world with unseen companions whose variety and extent dwarf that of the larger organisms of which we are more commonly aware.

The study of microorganisms must necessarily proceed in a way that takes into account their small size. For example, a single bacterium might weigh on the order of a picogram (10^{-12} gram). Such a small object can—just barely—be viewed with a light microscope, but resolution of such structural details as membranes, flagella, or ribosomes requires electron microscopy, with all of the specimen preparation and other technology it entails.

Microbial Cultures

Chemistry on such a small scale is extremely difficult to study and it makes excellent sense to work with large numbers of identical bacteria. Indeed, because bacteria and other microbes are often identified on the basis of their chemical activities, large numbers of cells are frequently required just to determine what the organisms are. Fortunately, this is entirely feasible: in the laboratory, a single bacterium (or other microbial cell) can, by cell division, rapidly give rise to a population containing grams, or kilograms, of identical (or nearly identical) cells, whose composition or biochemical capabilities can then be investigated. Such a population of identical cells, maintained in the laboratory under controlled conditions, is called a culture.

This culture strategy works because many microbes grow rapidly—much more rapidly than they mutate—leading to the creation of a large, homogenous population. Many prokaryotes are noted for particularly rapid growth, with, for example, populations of Escherichia coli, growing under just the right conditions, doubling as often as every eighteen minutes. By contrast, a human population might, but shouldn't, do the same thing in eighteen years.

Cultures are populations of microorganisms that may be maintained in the laboratory, although one can speak of cultures occurring in nature as well. Pure cultures are populations containing only a single species. Much of what we know about the biology of the *Archaea* and also *Eubacteria* has come from studies of pure cultures and the method has been extended with great success to the culture of eukaryotic cells, including human ones. In fact, it is not hyperbole to assert that pure culture methodology is so important in the study of cells that it very nearly defines modern microbiology.

Because much of our knowledge about Archaea has been obtained by means of pure cultures, it is useful to tell how microbiologists go about obtaining pure archaeal cultures and how they maintain them in the laboratory. The act of obtaining a pure culture from a real-world sample, such as a volume of seawater or a handful of soil, is termed the isolation of that organism. And because many *Archaea* can endure extremes of heat, salinity, and so forth their isolation can be greatly simplified by elimination of less hardy competition by imposition of such rigorous conditions.

The Idea of a Pure Culture

Pure cultures—populations of individuals belonging to a single species—must be kept under conditions such that other organisms cannot contaminate them. The precautions one takes to accomplish this are lumped together under the term "aseptic technique" and include sterilizing the flasks, nutrient solutions, and implements, such as pipets, to be used. Sterilizing means killing any organisms that are present, and is accomplished with heat, toxic gases, or ionizing radiation. Heat is often imposed in the form of steam under pressure, as in a pressure cooker, known in the laboratory as an autoclave. An autoclave is effective since, under pressure, the temperature can far exceed the ordinary boiling point of water. Solutions of chemicals may also be sterilized by filtration, the physical removal of any organisms that are present.

The chemical mixture in which a culture grows, and which provides nutrients to support it, is called a medium. Media may be liquid, with the nutrients in solution. Or they may be solid, as in the case of a nutrient liquid to which a solidifying agent, such as agar, has been added while the medium is hot. Agar, used in this fashion, produces a medium with the consistency of a jelly and, in most instances, organisms grow, either as discrete colonies or a continuous "lawn" on the medium surface. Solid media are often dispensed in the molten state into flat dishes with overlapping covers. These are called petri dishes or just plates. One pours a solid medium into a plate and one plates (i.e., spreads on a solid medium surface) a sample of bacteria.

Moreover, media can be "defined" or "complex." Defined media are completely specified as to their chemical composition, that is the medium is prepared by weighing out known quantities of pure chemical ingredients. In contrast, media may be "complex," which simply means not defined. Yeast extract is an example of a complex medium, as are a variety of extracts from meats. In general, growth in a defined medium is to be preferred from the point of view of reproducibility and control. But growth in a complex medium is often more rapid and, in some instances, there is no choice: successful defined media have not been developed for many *Archaea*.

Another reason pure cultures are so valuable in the study of microorganisms is their uniformity. They consist of many identical organisms, so that the properties of the organisms can be studied in bulk. And a culture usually contains an enormous number of individuals. For example, a one-liter culture of a microorganism that has been grown to a visible density might contain 10^{12} cells, definitely a more than adequate statistical sample. Attributes of the bulk culture are statistically valid representations of the situation in individual cells. A related benefit of being able to obtain large volumes of identical cells is the ability to carry out chemical analysis on a scale suitable for precision and convenience. Chemical determinations using a cell, or a few cells, are often difficult and inaccurate; large volumes for "macro" determinations are normally much more appropriate.

Isolating Pure Cultures

A central strategy of microbiology, called enrichment culturing, consists of placing an environmental sample, with its complex mixture of microorganisms, under conditions that favor the growth of one, or a few, kinds. Such selectivity often resides in the composition of the growth medium because different organisms exhibit strikingly different nutritional preferences. Likewise, oxygen is an absolute requirement for growth of many organisms, but extremely harmful to others, so that its presence or absence can exert strong selection. For example, even low concentrations of oxygen are fatally toxic to methanogenic *Archaea*, and the gas must be rigorously excluded from their growth media. Selectivity can also be provided by physical conditions, of which temperature and pressure are especially important factors in the case of the *Archaea*.

Mention should be made of the choice of inocula, the environmental source of the organisms being isolated. There is a well-founded prejudice in microbiology that "everything is found everywhere"—that almost any inoculum can serve as a source for almost any organism. This view considers that the selection of organisms finally isolated reflects only the culture conditions. This is largely the case, but it does improve one's chance of isolating a photosynthetic organism to use inocula from environments exposed to light. Likewise, it is only sensible to seek thermophilic *Archaea* in hot-spring water or volcanic soils. One does succeed in isolating organisms from unlikely locations, but this is not necessarily how one optimizes one's chances. Recall that many microorganisms occurring in nature cannot be cultured under known laboratory conditions. Thus, one succeeds only with a small fraction of the organisms present, and enhances the success rate by choosing an appropriate inoculum has considerable merit.

The strategy for obtaining pure cultures from a messy inoculum that might contain thousands of resident species usually consists of two phases. First, the use of selective growth conditions narrows the field. If a particular sugar is the only carbon source provided in the medium, then only those species capable of metabolizing it will grow. Likewise, if the pH is set at the very acidic value of two, or if the temperature is held at 80 degrees (or both), then organisms unable to grow under these extreme conditions will be eliminated. Sometimes, in the case of particularly rugged *Archaea*, such an enrichment procedure yields, in a single step, a pure culture. However, most often, this first phase of isolation leaves one with a mixture of different microorganisms, no longer numbered in the thousands, but still a mixture.

The task of obtaining a pure culture then consists of separating the individuals in this mixture and growing any, or all, as pure cultures. This is phase two of the isolation process and usually amounts to the physical separation of the culture into individual cells, each of which can, by repeated division, give rise to a pure culture. Such physical separation can be accomplished by using a microscope together with a sterile micropipet to obtain single cells; these can then be transferred to a small volume of sterile medium. This method leads to a pure culture but, owing to the small size of the cells in question, can be technically difficult.

An easier way to accomplish the separation is to place a dilute sample of the mixed-organism culture onto the surface of a petri dish containing solid growth medium. If the culture is sufficiently dilute, then individual cells will give rise to isolated colonies, each a clone consisting of all the descendants of an initial cell that found itself on the solid medium. These colonies are, by definition, pure cultures and the colonies may be removed sterilely and introduced to a suitable liquid medium.

Dilution of the initial culture is required so that individual cells can be grown in isolation from one another, leading to individual clones. This can be accomplished in two ways. Of course, one can simply dilute a culture with a large volume of sterile medium, but microbiologists routinely employ a much easier technique called plate streaking. Here, one uses a small metal loop, which has previously been sterilized in a flame, to transfer a drop of the culture to the surface of the solid medium in a petri dish. This drop is then spread over the surface of the medium, streaking in such a pattern that, at the end, there are very few cells left. When the petri dish is incubated for a sufficient time, the part of the surface where the streaking began will have a continuous smear of growth, but there will be regions of the surface where isolated colonies are found. Cells from these colonies can then be transferred to other solid or liquid medium and should constitute pure cultures.

Growing "Extremophiles" in the Laboratory

Many *Archaea* hail from extreme habitats and are appropriately called extremophiles. Naturally, they must be cultured under appropriately extreme conditions. For example, those isolated from acidic hot springs must be grown at a low pH, without oxygen, and perhaps at high concentrations of sulfides. Such a regimen leads to practical problems. Growth media can be highly corrosive, especially at a high temperature, and cells must be grown in chemically inert vessels. Likewise, growth on solid medium presents its own difficulties, as ordinary agar melts at temperatures that are often required. Fortunately, alternative solidifying agents with higher melting points have become available: for example, Gelrite, a naturally occurring gum is commonly used in place of agar for this purpose.

Many *Archaea* (and all methanogens) are strictly anaerobic, and even small quantities of oxygen can completely inhibit growth. For this reason, such organisms are grown only under an atmosphere of oxygen-free nitrogen or argon and, in order to remove the last traces of oxygen, it is often necessary to include oxygen-trapping chemicals in the growth medium itself.

Storing Cultures

Pure cultures can be maintained for years by sequential transfer into sterile medium and can provide any mass of cells required for use in investigations. Laboratories that work with particular organisms often maintain them by sequential transfer of small cultures, from which inocula can be taken to start larger cultures to be used for some particular application. Or if maintaining a culture in this serial fashion over a long period becomes a nuisance, then there are ways of storing cultures in inactive, but viable (i.e., recoverable) form. Cultures of many microorganisms can simply be frozen and thawed when required. In some instances, they remain viable at the temperature of the freezing compartment of an ordinary refrigerator; in other cases, they must be frozen and maintained at a temperature as low as that of liquid nitrogen (-196 degrees.) An alternative method for maintaining cultures in inactive form for long periods of time is lyophilization, or freeze-drying. Here, the culture is frozen, and water removed directly from the ice under a vacuum; this procedure minimizes cellular damage and lyophilized cells can be revived by introduction to a suitable sterile medium after many years.

In this connection, repositories have been established in which microorganisms are stored under appropriate conditions and where, for a fee, one may obtain them for research, education, or commerce. Two such repositories that maintain wide selections of *Archaea* are the American Type Culture Collection in the United States and the DSM (Deutsche Sammlung von Mikroorganismen und Zellkulturen GmbH) in Germany.

Viability

Microorganisms, both *Archaea* and *Eubacteria*, are often quite tough and can survive for very long periods, even under conditions that do not support growth. In such a circumstance we say that they have remained viable—able to resume growth when the situation permits. Archaeologists have been known to register personal concerns about potential disease-causing bacteria from 4000-year-old Egyptian mummies and, at least in principle, the concern is valid. It is entirely conceivable that pathogenic bacteria could survive that long, but not as certain that they could tolerate the dehydration associated with the mummification process.

The ability of cells to remain viable, but not growing, for extended periods can greatly favor their spread from one location to others. For example, the extremely thermophilic archaeon *Thermococcus peptonophilus*, isolated from deep-sea vents, grows optimally at 85 degrees but, when not growing, rapidly loses viability at that temperature. (It loses viability much more rapidly at 98 degrees, a temperature that still permits growth). In contrast, this organism survives very well at 4 degrees, especially if oxygen is available. It happens that 4 degrees is a characteristic temperature for deep seawater so that long-term viability at that temperature may constitute a mechanism for dispersal of *T. peptonophilus* from one vent site to another.

Isolation of *Sulfolobus Acidocaldarius*

The isolation of *Sulfolobus acidocaldarius* was reported in 1972 by Thomas D. Brock and three associates from the University of Wisconsin. As we learned earlier, Professor Brock is a world authority on thermophilic bacteria, and many of them were first isolated and investigated in his laboratory. He is also a microbiologist of uncommon breadth and was, for many years, senior author of a highly regarded textbook of general microbiology, listed in the Additional Reading section under the name of its present senior author, M. T. Madigan. Recall also that, at the time when *Sulfolobus* was first isolated, the *Archaea* had not yet been discovered, and this new genus was considered bacterial.

Here is a recipe for isolating *Sulfolobus*. Start with a soil or water sample (inoculum) from acid hot springs or their environs. Successful inocula have come from such locations in the United States, Italy, El Salvador, and Dominica. Add to the inoculum 0.1 percent acidified yeast extract solution and in-cubate (i.e., let the mixture sit) for several days at a tempera-ture of 70 degrees Centigrade. This is 158 degrees Fahrenheit, and unpleasantly hot to the touch. Yeast extract is a commer-cial product and contains just about any molecule found in an organism, including an extensive menu of sugars, amino acids, and vitamins. It can thus be assumed to contain just about all the chemicals that most organisms require. Prior to use, the yeast extract solution is brought to an acidic pH by addition of an acid like sulfuric acid. In most isolations, the pH of the medium is two, but can be as low as one. This is *very* acidic, indicating a very high hydrogen ion concentration. (Recall that the pH scale is logarithmic: a pH of one is ten times as acidic as a pH of two. Conversely, a pH of ten denotes a very low hydrogen ion concentration and a very basic solution.)

After several days, the initially clear culture medium is seen to have turned cloudy, and microscopic examination reveals it to have become a dense, and somewhat smelly, suspension of small cells. Although the rich chemical composition of the me-dium would support the growth of any number of different microorganisms, in fact, only one kind actually grows. The high temperature and the acidic pH effectively eliminate the com-petition. Indeed, the culture of *Sulfolobus acidocaldarius* ob-tained in this fashion turns out to be pure—only containing that single species. Thus, the ability of many *Archaea* to grow under uncommonly rigorous conditions can make them quite simple to isolate.

The population of *Sulfolobus* may then be maintained as a pure culture by subsequent transfer of a small volume of it into a larger volume of the same medium, acidified 0.1 percent yeast extract. One should never advocate carelessness, but with an organism like this, maintaining a pure culture is a rather simple affair. Normally, microbiologists must be extremely careful not to introduce foreign organisms into a pure culture and are scrupulous about the sterility of added medium, flasks, and pipets. In this instance, there are few prospective contam-inants able to grow at a pH of two or at 70 degrees. In that respect, at least, *Sulfolobus* is a good organism for neophytes.

Finally, the long-term culture of *Sulfolobus* entails an inter-esting quirk that can be both a minor bother but also instruc-tive concerning the biology of the organism. When a culture in liquid medium has stopped growing, it is customary to place a small inoculum of it in fresh, sterile medium so that growth can continue. With most organisms, the amount transferred to the new medium doesn't much matter: cells divide by doubling, so that even a few cells quickly give rise to an enormous population. But, in the case of *Sulfolobus*, it does matter. If the inoculum contains too few cells, the cells won't grow and the culture can be lost. Thus, there is usually a minimum cell density required for growth, below which the cells are not viable. The explanation for this behavior appears to be the leakiness of the plasma membranes that surround *Sulfolobus* cells, especially at the high temperatures at which these thermophiles grow. In other words, they tend to leak small molecules that are required for growth, and a large external volume favors that leak. The remedy is to use a large inoculum. Thus, even with the leak, the concentrations of the molecules in question remain high enough to keep the *Sulfolobus* cells, as microbiologists like to put it, "happy."

When cells are grown in culture, a measure of their "happiness" is their growth rate. Clearly, cells grow most rapidly under their optimal physical and nutritional conditions and effort is often expended to determine just what those conditions may be. For one thing, the relationship between cells' environment and growth rate can provide important information about their biochemistry and nutrition. And when microbial cells find themselves in favorable circumstances, their growth can be prodigious. Remember that a population of cells of the intestinal bacterium *Escherichia coli*, growing at 37 degrees in a rich medium, can double in about eighteen minutes. By way of comparison, the doubling time for *Sulfolobus acidocaldarius* at 79 degrees in a rich medium has been reported to be 213 minutes—about three-and-one-half hours. Other thermophiles appear to take about twice as long to double, but in no case can one ever be certain that conditions are really optimal.

Isolating Halophilic *Archaea*

Not many microorganisms can grow in concentrated brines, solutions containing from 10 to over 30 percent total salts. For this reason, those that can—the halophiles—are relatively easy to isolate in pure form. Indeed, the higher the salt concentration that they can endure, the greater the ease of their purification, as competition is progressively minimized. Organisms that can tolerate salinity in excess of 25 percent are often called hyperhalophiles; some of these are even able to grow in saturated sodium chloride, which at ordinary environmental temperatures has a concentration of about 32 percent. A few animals (including brine shrimp), no plants, and a mere handful of microorganisms, including the alga *Dunaliella*, are the only other forms of life that halophilic *Archaea* encounter in such an environment.

Much of the Earth's surface is covered with dilute brine called seawater, but extremely saline environments are quite uncommon. To gain a sense of what is meant by "extremely," consider that the Great Salt Lake in Utah is about ten times as salty as ordinary seawater—its sodium chloride concentration is ten times as great. Locations that are saline enough to be likely sources of halophilic *Archaea* include bodies of water in situations where evaporation, together with low rates of freshwater influx, lead to high salt levels. The Great Salt Lake, the Dead Sea, and various lakes in Africa and central Asia are familiar examples. There are also instances of natural saline lakes that owe their elevated salt concentration largely to factors other than high rates of evaporation. Thus, parts of the Antarctic continent are dotted with lakes exhibiting a wide range of salinity and specific salt composition. In such instances, the salinity is increased, in part, by freezing, which concentrates salts in the liquid phase, followed by evaporation of pure water from the ice phase. The salt composition may be variably altered by selective precipitation of different ions when the water is cooled. Halophilic *Eubacteria* and *Archaea* have been isolated from such waters.

Halophiles have also been isolated from dry salt deposits where salt is mined commercially. It appears that such deposits crystallize from natural brines that have been buried by geological processes. For example, living halophilic *Archaea* and *Eubacteria* have been isolated from English salt mines, some of which are 12000 meters below the surface and were formed in the Permian period, over 200 million years ago. As the microbiologists in question exercised great care to avoid surface contamination, one might be tempted to conclude that the microorganisms isolated from such mines are relics of populations that inhabited ancient salt lakes. However, the situation is not that clear: deep groundwater is turning out to be a rich source of microorganisms, some of which appear to be transported underground over great distances. Therefore, the supposed antiquity of these halophiles remains in doubt.

The discovery of halophiles in salt mines may be a case of human activity uncovering an ancient microbial community. In other instances, human activities can actually produce environments that are suitable for the growth of halophiles. For example, table "sea salt" is produced commercially by evaporation of seawater from flat basins called salterns, which are features of the littoral landscape in portions of the Mediterranean and other sunny ocean margins. When the salt concentration in these salterns has become sufficiently elevated by evaporation, halophiles often grow rapidly and color the water red to purple with their cellular pigments. In similar fashion, brines used in industrial processes and highly salted fish can acquire a noticeable reddish color from halophile contamination. Fortunately, none of the halophilic *Archaea* have proved to cause disease, so that effects of the contamination are merely aesthetic.

Natural high-salt environments differ significantly in their salt composition, depending on the mineralogy of the region. Thus, the Great Salt Lake contains water with a balance of salts similar to seawater, but, as we saw, with ten-fold higher concentrations. In both cases, sodium is the predominant cation, chloride the predominant anion. In contrast, the Dead Sea is much lower in sodium and correspondingly higher in magnesium, whereas so-called soda lakes, found in northern Africa and parts of the American West, are rich in carbonate and, for that reason, exhibit high pHs. Each of these different salt lake environments contains their own characteristic halophilic *Archaea*. For example, soda lakes are places to look for members of the genera *Natronobacterium* and *Natronococcus.* (Recall meeting *Natronobacterium* in Chapter 1.) In contrast, the biblical Dead Sea is the home of halophiles, like *Halobacerium sodomense*, that require a high magnesium content.

Many halophilic *Archaea* were first isolated long before the archaeal domain had been established. Thus, in 1980, when F.

Rodriguez-Valera and associates isolated a species of *Halobacterium* from a salt evaporation pond near Alicante in southern Spain they considered it to be bacterial. But, in fact, it was a halophilic member of the *Archaea*. Its isolation was carried out in the following fashion. The inoculum consisted of water from the saltern to which they added the sugar, glucose, plus ten inorganic salts. The total salt concentration was almost 20 percent, with sodium chloride, magnesium sulfate, and magnesium chloride being the most concentrated components, in that order. The medium was at neutral pH, a pH of seven. Finally, the antibiotic penicillin was included because it had been observed to prevent growth of some mildly halophilic organisms that could otherwise have contaminated the culture. Thus, penicillin imposed an additional selective pressure on the culture.

This medium might seem complicated, but it is, at least, a defined one. Also, all the carbon comes from the single sugar, glucose—in contrast to many organisms, including many halophilic ones, in which literally dozens of carbon compounds are required for growth. It is really quite remarkable that any organism is able to get all its sustenance from such a simple collection of chemicals, all obtainable "off the shelf" from a chemical supply company.

After inoculation, the culture was incubated at 39 degrees. Growth, as indicated by a pink cloudiness, was observed after two or three days. Samples from this enrichment culture were then spread on solid medium, and individual colonies selected as examples of pure cultures. At this stage, the colonies that grew were uniform and probably identical samplings of the same species. When viewed with a light microscope, cells were variable in shape, ranging from ordinary rods to disk and spheres. And not being equipped with flagella, as quite a few *Archaea* are, they were not motile. The authors noted the similarity of their organism to the previously isolated *Halobacterium volcanii*, but pointed out that the former differed because it had no complex nutritional requirements, growing, as we saw, with glucose as sole carbon source. They didn't propose their organism as a new species, but they well might have.

Isolating Methanogens

Marine microbiologists have recently taken to using submersibles—small research submarines—to obtain samples from hydrothermal vents; their voyages have led to the isolation of numerous *Archaea*. Such sampling has the virtue of allowing visual observation as well as study of physical and chemical properties of water where the organisms reside. An example of the results of such exploration is the isolation, by Karl Stetter, Carl Woese, and colleagues, of a new species of methanogenic *Archaea*, which they named *Methanococcus igneus.*

Carl Woese is, by now, familiar as the discoverer of the unique (domain) status of the *Archaea*. Karl Stetter is also a notable pioneer in the biology of the *Archaea*, having discovered many new archaeal species and made singular contributions to understanding their biochemistry, molecular biology, and place in the living world. The name of this organism indicates that it is methanogenic, that its cells are spherical (i.e., coccus-shaped), and that it is found in an extremely hot place. To be specific, it was obtained at a depth of about 100 meters in a hydrothermal system north of Iceland, that hotbed of geothermal activity. The water temperature was approximately 90 degrees.

Here is how the organism was isolated. The research submersible *Geo* was used to survey the vents on a submarine ridge north of Iceland. Samples of black sediment, together with venting water, were sucked into sealed glass vessels located on the outside of the vehicle, using what the investigators called a slurp gun. The object of this particular dive was isolation of methanogens, organisms that are strictly anaerobic and indeed can be killed by oxygen. Therefore, it was necessary to exclude oxygen during the isolation process, and reducing agents, sodium sulfide and sodium dithionite, were immediately injected into the glass vessels. Sealed samples were subsequently transported to the laboratory and placed in an anaerobic culture medium that contained only a mixture of inorganic salts—a "mineral medium." The culture was then incubated at 85 degrees under an atmosphere of hydrogen and carbon dioxide; oxygen was rigorously excluded from the flask. The only source of carbon in the medium was carbon dioxide, so that any growth that occurred would be "autotrophic." Autotrophic organisms, of which green plants are a good example, manufacture all their carbon compounds ultimately from carbon dioxide. In contrast, heterotrophic organisms depend on organic carbon sources—compounds like sugars, amino acids, and so on—not being able to assimilate carbon dioxide. (We humans are "heterotrophic.")

After one day, the culture medium was observed to have become cloudy and spherical cells were seen with the light microscope. The cells were examined in a fluorescence microscope and observed to emit a blue-green fluorescence characteristic of a cellular component found only in methanogenic *Archaea*. At the same time, large quantities of methane, CH_4, were produced by the culture; the methane was formed from carbon dioxide and hydrogen. Clearly, the new organisms were methanogens.

The isolation process continued with the plating of the culture of the same medium solidified with agar. These plates were incubated at 75 degrees under the carbon dioxide-hydrogen atmosphere and, after three days, small yellow colonies were observed. These, of course, were pure cultures and, reintroduced into liquid medium, could be used to grow large amounts of cells for further studies. In this case, further studies included 16S rRNA sequence analysis that supported the conclusion that the organism was a newly discovered methanogen. The organism, having been isolated and grown at a high temperature, is thermophilic and is an unusual methanogen in its ability to grow in an extremely simple mineral medium, without organic carbon sources, vitamins, or other growth factors.

Growing Large Quantities of Archaeal Cells

Knowledge about the chemical composition and physiology of microorganisms generally requires the growth and harvesting of large amounts of cellular material. Pure culture techniques make this "farming" possible, and it goes without saying that care must be taken to avoid contamination with unwanted species. The goal is to obtain kilogram quantities of cells of a single species, cells grown under completely known and reproducible conditions.

One usually begins with a small culture that has been in some sort of long-term storage. We saw that prokaryotic cells are often quite durable and can be kept alive for long periods of time, either in the frozen or the lyophilized state. In either case, if the stored organisms are viable—that is, capable of revival—then they can be started up by transfer to a small volume of a suitable medium and grown under suitable conditions.

Subsequently, large-scale growth can be carried out under conditions dictated by the nature of the organism. Thus, if the cells are to be grown in the presence of oxygen, cultures can be placed in large, partially filled flasks, and shaken vigorously on a rotating platform in order to disperse air into the culture. Or the cells may be grown in a fermentor, a large sealed vessel equipped with a stirrer and ports that allow addition of liquids and gas, and accommodate pH and oxygen electrodes. Fermentors are particularly useful for the large-scale growth of anaerobic organisms like methanogens. In this instance, "large-scale" denotes volumes ranging from a few to many hundreds of liters. Finally, there is a type of fermentor called a chemostat that is designed so that fresh, sterile medium can be added continuously to a growing population of cells. As new medium is added, an equal volume of medium containing organisms leaves via an overflow and can be collected. Thus, the chemostat can continuously produce cells that have been grown under constant and precisely regulated conditions, assuring excellent quality control in the cells produced.

The Harvest

Cells, produced by whatever method, are then harvested—that is, separated from the medium in which they were grown. This is usually done by centrifugation, and the last traces of spent medium and waste products are removed from the cells by repeated resuspension in liquid followed by centrifugation. The cells, "washed" in this fashion, may then be studied directly as intact cells or, in the case of many biochemical studies, can be fragmented and their various cell parts separated. Because *Archaea*, and prokaryotic cells in general, are often very tough, scientists frequently must employ extreme methods to disrupt them. These include passage of cell suspensions through needle valves under extremely high pressure or introduction of ultrasound into a closed volume; both treatments produce large shear forces in and around the cells. Unbroken cells and large fragments may then be removed by centrifugation. If subsequent studies require it, the resultant cell extract may be separated into fractions such as soluble enzymes, ribosomes, and membranes, also by means of centrifugation.

Archaeal Portraits

Archaea include some of the most bizarre cells in the entire living world. Some of them have the form of isosceles triangles, while others are extended threads, one hundred times as long as they are thick. Still others are precise squares and resemble postage stamps. And quite a few are knobby, irregularly covered with small lobes. On the other hand, many *Archaea* appear quite unremarkable, taking the form of nondescript rods or spheres, just like the majority of *Eubacteria*. But, in all cases, the detailed cellular architecture of *Archaea* is unique. For example, most *Archaea* have cell walls with chemical compositions unlike any other cell walls in the living world. Likewise, their cytoplasm contains ribosomes that differ in shape and composition from all others. And they all are enclosed by membranes with unique chemical composition and overall appearance, quite unlike those belonging to any other organism that we know about.

Archaea, as prokaryotes, have a relatively minimal cell plan. By contrast, the eukaryotic blueprint is more baroque, with lots of complexity in the form of interior membrane systems. Similarly, eukaryotic DNA, unlike prokaryotic DNA, is decorated with nucleosomes and normally packaged in the form of complicated objects called chromosomes. The occurrence of chromosomes, and also the mitotic apparatus that partitions them during cell division, are consequences of the much larger genome of eukaryotic cells. The packaging and other arrangements aid in managing the large amount of genetic information—rendering it accessible and ensuring equal partition when cells divide. (It is good that DNA is divided among chromosomes in the same sense that it is fortunate that libraries contain individual volumes, not just one large volume.) Of course, the large amount of DNA in eukaryotic cells is itself a consequence of their much greater complexity, when compared to prokaryotes.

Prokaryotic cells, with their relative simplicity, are designed for growth—often opportunistic growth under a wide variety of conditions. With such a focus on growth and adaptability, there is little room for frills. The much more complex, and to the human eye, more elegant, architecture of eukaryotic cells is designed for growth too, but much of the complexity is there to allow for greater subtlety of regulation, especially in light of the much larger amounts of genetic information that such cells generally contain. Indeed, most of the important differences between the two categories of cells reflect, in one way or another, the different sizes and patterns of organization of their genomes.

These large eukaryotic genomes, together with opportunities for regulation of gene expression that result from the greater complexity of eukaryotes, lead to the possibility of multicellular organisms composed of a variety of cells. Prokaryotic cells also sometimes occur as aggregates but, in general, the cells are alike. Thus, a complex organism that is composed of different cell types must be provided with ordered ways of turning on and off different genes in the different cells—a process called differentiation. This process requires a rich genome (as many genes will be inactive in any particular cell) and a patterned program of regulation. These appear much more feasible in the eukaryotic cell format and, indeed, it can be suggested that the rise and evolution of eukaryotic organisms were promoted by the adaptive advantages of being multicellular.

In illustrating the contrast between the two cell formats, one could do worse than paraphrase a fragment from the poet Archilochus concerning the hedgehog and the fox. The prokaryote is the hedgehog, who knows ''one big thing'' (opportunistic growth). The eukaryote is the fox: it ''knows many things,'' and knowing many things in the case of a cell entails subtlety of regulation and consequent complexity of structure. Let us examine this great dichotomy of cell architecture more fully before turning to the *Archaea*, keeping in mind that *Archaea* remained undiscovered, as a group, for such a very long time because of their structural similarity to ordinary bacteria.

Prokaryotic Architecture

When the distinction between prokaryotic and eukaryotic was first established, definitions of the former were mostly negative in character: prokaryotic cells were known for what they lacked. For instance, prokaryotic cells lack a membrane-enveloped nucleus, which is a defining feature of an eukaryotic cell. Likewise, eukaryotic cells are known for their intracellular membrane systems, which often include such organelles as mitochondria, chloroplasts (both engaged in energy transfer) as well as endoplasmic reticulum and Golgi complex (both involved with protein synthesis and packaging). Prokaryotic cells generally lack these organelles. Some of the membrane systems, such as mitochondria and chloroplasts, enjoy considerable autonomy in the sense that they contain genetic material, make some of their own proteins (or parts thereof), and have a partially independent evolutionary history. In contrast, prokaryotic cells have modest intracellular membrane arrays, if any, and those that occur never exhibit autonomy of the sort described.

If it appears somewhat awkward to define prokaryotic cells by what they are missing, fortunately this isn't completely necessary. For one thing, all cells contain ribosomes, the locations of protein synthesis, in their cytoplasm. Based on size, there are two kinds of ribosome in the living world: the lighter 70S ribosome, found in all prokaryotic cells, and the heavier 80S ribosomes, found in all eukaryotic cells. Thus, the two cell types are also characterized by the sizes of their ribosomes.

Prokaryotic Cell Walls

The outermost layer of most prokaryotic cells is a cell-wall complex, consisting of a rigid wall and often a softer glycocalyx.

("Calyx" denotes vessel—an object that contains something.) The prefix "glyco" denotes sugar, and the glycocalyx is composed, at least in part, of glycoproteins—that is, proteins with sugar molecules attached. The cell-wall complex protects the cell and allows nonspecific adhesion to other cells or inanimate objects. The rigid wall is also an important line of mechanical defense and regularly gives a cell its characteristic shape.

The large majority of prokaryotic cells that have a cell wall are extremely tough little creatures that strongly resist mechanical deformation or breakage. Indeed, biochemists often have great difficulty disrupting prokaryotic cells in order to study their interior. This toughness of prokaryotes reflects the physical strength of their wall as well as a force due to osmosis— the uptake of water. Think of the cell wall as a wicker basket in which a balloon has been inflated so that it exerts pressure from the inside. Such a basket is very rigid and resistant to mechanical damage. Thus does the prokaryotic cell (and eukaryotic cell that possesses a cell wall) gain strength from a flexible plasma membrane pressing against a rigid cell wall.

The explanation for that pressure is osmosis, a process by which water enters a cell until an equilibrium is reached. Water, like other compounds, tends to cross membranes in the direction of the concentration gradient, moving from a higher to a lower concentration. This movement is passive—like a ball rolling down a hill—it happens spontaneously so long as the gradient persists.

The osmotic water concentration gradient is the result of a higher concentration of dissolved protein, and other compounds, inside a cell than on the outside. Because these dissolved materials take up space, the concentration of water is necessarily much lower inside the cell. The concentration difference causes the movement of water down the concentration gradient, into the cell. The cell wall imposes a limit to cell volume and the osmotic movement of water exerts a pressure, plastering, as we saw, the membrane against the wall. This pressure increases until it exactly opposes the inward flux of water; that is, equilibrium is reached. Thus, the mechanical strength of a prokaryotic cell is largely a consequence of the force exerted by osmotic pressure against the inner surface of the wall.

Cell walls provide a decisive way to tell *Archaea* and *Eubacteria* apart. Walls from all *Eubacteria* contain a major component called peptidoglycan or murein; this does not occur in *Archaea*. Some methanogens do contain a modified peptidoglycan, or pseudomurein Other *Archaea* have only a protein coat, with no peptidoglycan derivatives at all.

When microbiologists identify prokaryotic organisms, they are aided by a simple cell-wall staining technique, the Gram stain, which sorts prokaryotes into two categories. Organisms that are stained are termed Gram-positive; those that don't are Gram-negative. Both prokaryotic domains, *Archaea* and *Eubacteria*, include Gram-positive and negative members, although most archaeal cells are Gram-negative. For example, *Sulfolobus, Methanococcus,* and *Methanosarcina* are Gram-negative, while at least one species of *Methanobacterium* is Gram-positive.

Appendages from Cell Surfaces

Many prokaryotic cells are decorated with small protein fibers called pili. These are quite common in *Eubacteria* and relatively rare in *Archaea*. Pili extend from the cell surface and are present to facilitate cell adhesion (to other cells or other objects) and also play a role in cell recognition, for example, allowing some bacteria to bind preferentially to other cells of an appropriate mating type.

A second sort of cellular appendage endows cells with the power of motion. Flagella occur in a variety of *Eubacteria* and *Archaea*, enabling them to swim through the surrounding water. The flagellum is a stiff, helical fiber composed of protein subunits called flagellins. A flagellum is itself passive and driven to rotate by a "motor" located at its base. The motor is also fashioned from proteins and has an intricate structure with fixed and rotating rings that would do justice to a particularly inventive mechanical engineer. Like all motors, this one converts some sort of fuel into physical motion. In this instance, the motor runs on direct current, a current of positively charged ions (or cations). Usually, hydrogen ion (H^+) is the ion of choice but, in some cases, it can be sodium ion (Na^+) instead. The archaeal flagellum motor has been studied most extensively in halophiles such as *Halobacterium*, where the identity of the ion current that drives it is still being settled.

Just as flagella are polymers of individual protein subunits, the flagellins, so are pili composed of pilins. The amino acid sequences of both kinds of subunit are known for a number of *Archaea* and *Eubacteria* and, in several instances, the locations of the genes have been pinpointed on the organisms' chromosomes. For instance, four flagellin genes for *Methanococcus voltea* are located in tandem on the cell's chromosome, so that their expression can be regulated as a group.

The nucleotide sequences of these four genes are extremely similar; this is not particularly surprising as the gene products, the flagellins, are alike in structure and function. It is surprising, however, that these genes are not at all closely related to the flagellin genes of *Eubacteria* but instead are highly similar to eubacterial pilin genes. It seems that flagellar structure constitutes another profound difference between life's three domains: eubacterial flagella are made of flagellins that are distinct from pilins, archaeal flagella are made up of flagellins that are pilins, and, to finish the story, eukaryotic flagella are not made up flagellins at all. Indeed, eukaryotic flagella are totally unlike those of either prokaryotic group—a feature that draws another clear line between prokaryotic and eukaryotic. Eukaryotic flagella have a complex tubular structure, in contrast to being a simple strand of protein subunits, and the power of motion is intrinsic to the flagellum itself and not provided by an independent motor.

Membranes Are a (Mostly) Unifying Feature of All Life

All cells are surrounded by plasma membranes and cells often contain internal membranes as well. With the notable exception of the *Archaea*, all these membranes look alike when viewed with an electron microscope. With the same notable exception, all membranes are similar in their chemical composition, in the overall way that they are put together, and in their physical properties, such as elasticity and electrical conductance. Prior to the discovery of the *Archaea*, this overwhelming sameness of biological membranes was cited as evidence that life originated only once, with all organisms descending from a common ancestor. That conclusion remains valid: but perhaps better evidence is the universality of the genetic code, wherein the same nucleotide triplets specify the same amino acids in all organisms.

However, there is that one striking exception to the uniformity of membrane construction: the *Archaea* are enclosed in membranes that are different from those of any other creature. Their membranes are made of different chemicals and organized in a strikingly different way from other organisms. So unique and so characteristic is the composition of archaeal membranes that *Archaea* may be reliably identified by it. Indeed, membrane components serve as valuable identifiers for the presence of *Archaea* in ancient rocks and sediments. Such membranes have unique properties, often including impressive stability at high temperature. It is as if the *Archaea* sprang from an entirely different ancestor than the rest of the living world, but they undoubtedly didn't. In fact, they, and all other organisms, undoubtedly share a common ancestor, but there is an enormous evolutionary distance between the *Archaea* and the rest of the living world: the *Archaea*, as we have seen, are only distantly related to other forms of life. This dramatic dichotomy between the almost-universal membrane pattern and the archaeal variant leads to an obvious question: which membrane pattern is closest to that of the earliest organisms, the common precursors of both *Archaea* and *Eubacteria?* One also wonders if there is a connection between the often extreme environments of many *Archaea* and their idiosyncratic membrane compositions. To address these matters, we must first describe membrane architecture and function a little more explicitly.

All living matter is physically separated from the nonliving world by cellular membranes, so that all material and all information that enters cells necessarily crosses them. Thus, membranes modulate all interactions between cells and their surroundings. For instance, the identity of the solute carriers in a cell's external membrane determines which solutes are to be transported and which aren't. That choice of solutes, in turn, strongly influences the character of the cell's metabolism and all that depends on it.

The (Almost) Universal Pattern of Membrane Architecture

We saw that a comforting assertion of cell biology used to be that all cellular membranes are practically identical. In an electron micrograph, they all certainly look the same, whether they came from the cell of an elephant or a bacterium. Prior to around 1970, this perceived similarity of all membranes extended to the intimate details of their chemical composition. Molecules called phospholispids formed the heart of all membranes and to these invariably were bound protein molecules whose exact identity depended on the membrane's specific origin and function. The general pattern dictated by the phospholipids and their interaction with proteins was always pretty much the same.

However, this "universal" pattern—a phospholipid core studded with bound proteins—as found in virtually all other organisms, does not extend to the *Archaea*. Archaeal membranes are usually not based on phospholipids, and their overall organization is strikingly different. The differences are so dramatic, and the surprise of biologists was so great when they were discovered, that it is appropriate to speak of a revolution in membrane biology.

Phospholipids: The Greasy Heart of (Most) Membranes

"Lipid" is a polite word denoting fat. Lipids may be distinguished by low solubility in water and high solubility in organic solvents like gasoline or cleaning fluid. This matter of solubility is instructive: lipids generally fail to dissolve in water because water is a polar molecule, with an electrical charge difference between its ends, whereas most lipids are not. Lipids are nonpolar and they don't interact well with water and don't dissolve in it.

If ordinary lipids are strictly nonpolar and don't dissolve in water, phospholipids are divided into a polar part (that interacts with water) and a nonpolar part (that doesn't). This ambivalent polarity underlies phospholipid's role as membrane building blocks. Thus, two layers of phospholipids associate to form membranes: the nonpolar parts of each layer form a core, with the polar regions on both sides, facing the water. If a molecule, such as a nutrient, crosses the membrane to enter the cell, it must traverse the nonpolar interior of the membrane. Polar molecules don't do that readily so that membranes form a tight barrier to their transit. If such a molecule is to enter the cell—for example, to serve as a nutrient—there must exist a special mechanism for its transport, usually in the form of a specific transport protein.

The nonpolar part of a phospholipid usually consists of two fatty acid chains that are about as nonpolar as anything can be, and so are extremely insoluble in water. These chains are composed of from sixteen to twenty carbon atoms and a variable number of hydrogens. Being long chains composed of carbon and hydrogen atoms, the fatty acids strongly resemble hydrocarbons found in gasoline or fuel oil—also extremely nonpolar molecules that are quite insoluble in water. The polar part of a phospholipid contains glycerol as a sort of backbone, to which is attached a phosphate. To the phosphate, in turn, is attached a polar region called a head group. Different head groups occur in different kinds of phospholipids, giving them their names. For instance, if the head group is choline, the phospholipid is named phosphatidylcholine.

Perhaps here is a helpful way to think about the geometry of a phospholipid molecule: the polar region is your fist; the two nonpolar (fatty acyl) chains are two extended fingers. A membrane is then represented by your two hands, with the four extended fingers touching. The phospholipid molecule thus possesses a "split personality": one end (the fist) interacts well with water. The other end (the fingers) is very nonpolar and doesn't interact with water at all. It is this "amphipathic" property that enables phospholipids to form the membranes of most organisms, but not, generally, the *Archaea*. In archaeal cells, molecules other than phospholipids underlie the structures of their membranes, but it is the amphipathic character of these compounds that makes membrane formation possible.

A final distinction between archaeal and all other membranes is the nature of the linkage between the polar glycerol and nonpolar fatty acyl chains. Here again *Archaea* differ from other organisms, employing ether, rather than the much more ubiquitous ester links. It is enough to know that ether linkages are simpler than ester ones, involving three, instead of four, atoms. Perhaps because they are simpler, they are also more stable, which may partly account for the rugged character of archaeal membranes. In any event, this is another difference that definitively separates the *Archaea* and *Eubacteria*.

Amphipathic Molecules Form Membranes Spontaneously

When suspended in water, phospholipids and analogous archaeal lipids adopt an interesting solution to the dilemma of being part polar and part nonpolar. Thus, if one stirs some phospholipid vigorously in a bucket of water, one can observe the individual molecules lining up in double layers so that the fatty acid chains are directed toward other fatty acid chains and the polar ends are directed outward, toward surrounding water molecules. That is, the polar parts associate with polar, and the nonpolar parts with nonpolar, with the phospholipids automatically forming a bilayer. This is a molecular sandwich with the (nonpolar) butter on the inside. In the case of the phospholipid in the bucket, the phospholipid would turn out to have formed a suspension of spherical vesicles—little balloons of phospholipid, with walls composed of the bilayers that we have just described. Similarly, in real life, cells make membranes by simply manufacturing the phospholipids, after which the phospholipids do the rest, forming membranes spontaneously.

Of course, membranes are more than just lipids: they contain proteins that are often associated with solute transport or exhibit enzymic activity. The orientation of proteins in membranes is very specific, with perhaps a particular region of a protein being exposed on a particular side of the membrane. The correct emplacement of proteins in membranes generally occurs by complicated transport processes that are specific for particular proteins and particular cells, and we will not describe that here. But, in general, these processes often employ polarity, and polar and nonpolar regions of proteins often end by being in proximity to comparable regions of the lipid portion of the membrane.

No wonder that most cellular membranes look (and are) essentially the same: their architecture reflects universal chemical properties like polarity. And the same principles of polarity are at work in archaeal membranes, with nonpolar regions being in proximity to other nonpolar regions. However, we are about to learn that the identities of the molecular ingredients are actually quite different from the standard phospholipid case.

Which Evolved First (Ether or Ester Lipids)?

A compelling case can be made for the idea that the ether-based lipids found in the *Archaea* are extremely ancient and preceded ester-based lipids in the history of life. The present distribution of ether lipids says it all: significant amounts of ether lipids are found in all categories of *Archaea* as well as in the most ancient among the *Eubacteria*. Moreover, many *Archaea* and most *Eubacteria* that employ ether linkages are thermophilic. As the early Earth was a rather nasty place, with an atmosphere and ocean that were hot and perhaps acidic, the greater stability of ether lipids makes them good candidates for membranes in early cells.

Finally, the shapes of the nonpolar chains from archaeal membranes differ from those of all other organisms. In most organisms, the nonpolar chains are straight fatty acids, whereas the corresponding chains in *Archaea* are usually branched. In many instances, the archaeal chain exhibits a repeating pattern of branched, "isoprenoid," units, each containing five carbons. All *Archaea* have such branched chains, whereas all other creatures contain mostly straight chains. Clearly, this is a very deep-running distinction.

Lipid Bilayers Are Almost Universal (But Not Quite)

Until quite recently, everyone considered that *all* biological membranes are based on phospholipid bilayers. We now understand that archaeal membranes are exceptions because they commonly lack phospholipids, but they also often lack a bilayer format as well. Archaeal membranes are frequently constructed from single large molecules spanning the membrane, constituting a lipid monolayer. However, one basic principle of membrane architecture remains valid in all instances: every membrane—archaeal or otherwise—still shares the important feature of having a nonpolar region on the inside, more polar regions on the outside. The buttered sandwich model for membranes remains valid: the *Archaea* simply accomplish it differently, using a single molecule to span the membrane. That single molecule has a nonpolar region in its middle.

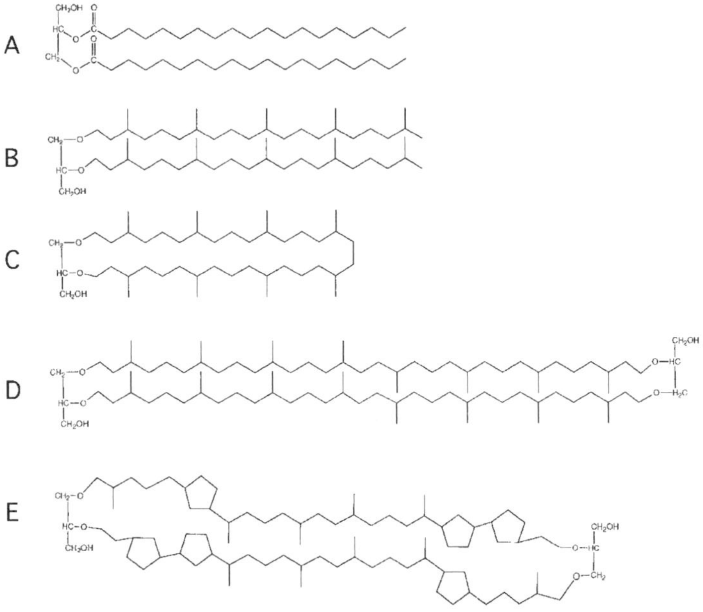

Figure 4.1 Backbones of some membrane lipids. **A** is that of an ordinary phospholipid. The phosphate is attached to the free hydroxyl (-OH) group. Note that the long, hydrophobic fatty acid chains—the zigzag lines—are connected to the three-carbon glycerol part by the ester linkage. **B** and **C** are archaeal lipids that span one-half of the membrane; **D** and **E** are archaeal and span the entire membrane; both have a glycerol at each end (at both surfaces of the membrane). All the archaeal lipids (**B–E**) contain either linkages between the glycerols and the hydrophobic portion. Notice that the hydrophobic parts of **E** contain five-membered rings. These constitute another unique feature of archaeal lipids.

We should also mention the sizes of membrane lipids. First, ordinary phospholipids usually contain fatty acids with from twelve to twenty carbon atoms; there usually are even numbers of carbons, owing to the chains being manufactured in two-carbon increments. In contrast, most archaeal lipids have branched (isoprenoid) fatty alcohols with from twenty to forty carbon atoms, depending on whether one, or two, of the chains spans the membrane. Such an isoprenoid twenty-carbon chain has been found in the membranes of all *Archaea* that have been studied, thus appearing to be a universal core lipid in the entire group of organisms.

Membrane Fluidity

Individual lipid molecules are able to move laterally in a membrane, a process described as fluidity. Fluidity affects all aspects of membrane function, and so must be tightly regulated in all living cells. Such movement is temperature dependent; membranes are more fluid at higher temperatures if all other factors remain the same. Clearly, thermophilic *Archaea* possess membranes adapted to the elevated temperatures at which they live. In particular, they respond to environmental temperature by modifying their membrane lipids, so that it is possible to make a reasoned judgment about a cell's environmental origin by looking at its lipid composition. For instance, many *Archaea* contain five-membered cyclopentane rings. There may be from one to four of these rings present in a single tetraether lipid, and the number reflects environmental temperature. The rings appear to decrease fluidity, thus compensating for the higher temperature. On the other hand, *Archaea* isolated from extremely cold environments (like Antarctica) compensate by increasing the number of double bonds in their lipids, a modification that increases fluidity.

Cytoplasm

The cytoplasm is everything inside a plasma membrane—the entire cellular content. In all cells, its chief ingredient is water, in which large amounts of proteins and other solutes are dissolved, with most of the proteins being enzymes. The cytoplasm also can contain internal membranes, storage granules of various insoluble chemicals, as well as other small objects with specialized functions, like ribosomes and gas vesicles. Some of these inclusions are characteristic of particular types of *Archaea* and *Eubacteria*. For instance, some cells from both domains contain granules of polymerized inorganic phosphate. This is an energy-storage strategy, as the individual phosphates are connected by the same "high-energy" bonds found in ATP. In addition, some *Archaea* and *Eubacteria* that transact metabolic business using sulfur store chunks of solid sulfur in their cytoplasm. Finally, certain *Eubacteria* and *Archaea* store gas within a protein envelope called a gas vesicle. These can provide a means of flotation in certain cyanobacteria, which are *Eubacteria*, as well as among halophilic *Archaea*. Gas vesicles also occur in the methanogens, contain carbon dioxide or methane, of which the first is a metabolic raw material, and the second, a product.

The Nucleoid

A prokaryotic cell, either archaeal or eubacterial, lacks a membrane-bound nucleus, a primary hallmark of the *Eukarya*. Eukaryotic cells contain most of their DNA in a membrane-bounded nucleus, so that genetic information storage is physically separated from its expression (i.e., protein synthesis). In contrast, prokaryotic DNA floats free in the cytoplasm and usually consists of a single, very large, circular molecule. This molecule is tightly wound up into a compact skein, the nucleoid; if it were not condensed in this way, it would be many times longer than the cell in which it resides. Sometimes the DNA molecule is attached at a single point to the plasma membrane, an arrangement thought to assist in cell division by pulling DNA replicas apart as the cell wall elongates.

Nucleoids, being rather small and indistinct, are somewhat difficult to observe in living cells, but addition of DNA-specific fluorescent dyes enables them to stand out as brightly fluorescing spots. When cells of *Sulfolobus acidocaldarius* are observed in this way, each cell is seen to contain one or two nucleoids, each irregular in shape, rather like the cells themselves. In rapidly growing cells, nucleoids tended to be more compact than in cells that were viable, but not growing. When cells divide, the nucleoids first separate into pairs, presumably a sequel to DNA replication, and then move apart so that, when the cell constricts to form two, there is a nucleoid in each.

Ribosomes

Ribosomes are small particles sprinkled about in the cytoplasm and are, as we have seen, sites of protein synthesis. Recall that all prokaryotic ribosomes differ in size from all eukaryotic ones—70S as compared to 80S—and in eukaryotic cells, ribo-

somes are usually bound to intracellular membranes. In prokaryotic cells, which usually lack such membranes, ribosomes
are freely distributed in the cytoplasm.

Although *Archaea* and *Eubacteria* contain ribosomes of the
same (prokaryotic) size, there are important differences. For
one thing, their shapes, as revealed by high-magnification electron microscopy, are noticeably different and, indeed, differ
characteristically among the various groups within the *Archaea*.
Also, archaeal ribosomes exhibit a different protein composition than those of bacteria and these differences turn out to
be reflected in the sensitivity of their protein synthesis to certain antibiotics. For example, eubacterial protein synthesis is
inhibited by streptomycin, erythromycin, and tetracycline,
which is why these compounds are useful antibacterial agents.
These antibiotics do not inhibit our own eukaryotic protein
synthesis, which is the other reason why they are successful in
treating human disease: they don't kill us. But these same antibiotics also fail to inhibit protein synthesis in the *Archaea*, suggesting commonality between protein synthesis in eukaryotic
and archaeal, but not eubacterial, ribosomes. This commonality extends to the action of diphtheria toxin, which affects protein synthesis in *Archaea* and *Eukarya*, but leaves bacteria unscathed. The toxin is, of course, the product of a eubacterium
and so it is reasonable that it shouldn't harm the sort of cell
that produces it. We will encounter diphtheria toxin later, in
the context of protein synthesis (Chapter 8).

Cell Skeleton

The shapes of prokaryotic cells seem to be determined largely
by the configuration of their cell walls, whereas eukaryotic cell
form is more a reflection of an internal structural framework,
the cytoskeleton. This cytoskeleton is composed of several
kinds of filaments that, in turn, are polymers of protein subunits, of which two common ones are actin and tubulin. Certain
specialized cytoskeletal filaments are able to contract, obtaining the energy to do so from ATP or GTP. Such contractility
is the basis for a variety of eukaryotic cell movement, including
the beating of cilia, the internal flow of cytoplasm, and the
motion of chromosomes during cell division (mitosis). For example, microtubules—made of tubulin monomers—form the

spindle apparatus that partitions chromosomes in dividing eukaryotic cells. Because mitosis, and the mitotic spindle apparatus, are hallmarks of the *Eukarya*, and because *Eukarya* must have evolved from prokaryotic precursors (no other candidates exist) it would be of considerable interest to discover related proteins in prokaryotic cells.

The absence of cytoskeletal proteins was long considered a defining feature of prokaryotic cells, but there is increasing evidence for the occurrence of homologous (structurally and evolutionarily) related proteins in both bacteria and *Archaea*. For instance, FtsZ is a protein required for cell division in species belonging to both prokaryotic domains. This protein is apparently associated with the formation of the external ring that pinches off the two products of cell division. The protein is structurally related to the eukaryotic tubulin, a subunit of the microtubules that play a central role in the motion of cilia and in drawing chromosomes apart during cell division. Not only are there regions with sequence homology between FtsZ and tubulin, but both bind the high-energy compound guanosine triphosphate (GTP), which, in both cases, may provide energy for molecular movement. Both proteins also have very similar three-dimensional structure, as revealed by X-ray and electron crystallography. Thus, it is reasonable that tubulins and FtsZ evolved from a common precursor and that this family of proteins traversed the prokaryotic-eukaryotic transition. Indeed, these proteins may have played a central role in that transition: eukaryotic cells acquired mitochondria, chloroplasts, and possibly much more, by engulfing prokaryotic precursors—the process called endosymbiosis. And there is reason to believe that proteins like tubulin and Ftsz must have played a role in the movement required for the engulfing process.

Thus far, genes encoding such archaeal tubulin precursors have been found only in methanogens and halophiles belonging to the kingdom *Euryarchaeota*. Also, in three out of four cases where the sequences are known, the FtsZ genes are duplicated, and comparison of the sequences tells us that the duplication event occurred early in archaeal evolution. Duplicated genes are particularly likely foci for subsequent evolution—in this case of tubulins—because one can be available for evolutionary experimentation while the other, unmutated, can provide continuity of function.

Cell Shape: Archaeal Variations on the Prokaryotic Theme

Archaeal cells conform to a typical prokaryotic plan, the minimalist architecture that they share with *Eubacteria*. However, within this prokaryotic framework, the *Archaea* tend to push the design limits, coming in shapes or almost perverse variety. Some exhibit an absolutely typical rod-shaped or spherical bacterial plan ("bacillus" or "coccus," respectively). Others are disk, spirals, or filaments, or exhibit amoeba-like irregularity, with variable protuberances. Still others give the impression of almost mineral-like geometry: cubes, triangles, or postage stamps. As an example of amorphous cells, consider the now-familiar genus *Sulfolobus*. As the name suggests, their cell surface in bumpy, or lobed. They are more or less spherical, but with irregular lobes projecting in a rather disorderly fashion. Other irregular cells are bumpy spheres (e.g., *Thermococcus*) or are vaguely rod-shaped, but of wildly variable diameter (e.g., *Pyrodictium*). This last organism is also noteworthy for its habit of manufacturing networks of hollow filaments, chemically similar to prokaryotic flagella. It uses these to adhere to sulfur crystals, forming a rather messy film. All these examples are in the thermophilic group of *Archaea*, but there are instances of highly irregular cell plans among the halophilic and methanogenic *Archaea* as well.

Strict regularity in bacterial cells reflects an orderly procedure for the manufacture of the cell wall. Likewise, the details of cell-wall assembly determine whether the shape will be spherical, rod-like, or otherwise. Of lobed (or otherwise irregular) cells, one can only say that variable cell-wall synthesis produces such extravagances of asymmetry. But there is one group among the *Archaea* that carries the disorderly tendency one step further: members of the genus *Thermoplasma* don't make a cell wall at all. They are spherical, with variable diameters, ranging anywhere from one-fifth to five micrometers. One might add that the failure of this organism to surround itself with a protective cell wall is particularly enigmatic, given its site of origin. Most cultures of *Thermoplasma* have been isolated from hot coal mine wastes that derive their heat from spontaneous combustion. It seems that *Thermoplasma* cells protect themselves from the evident rigors of this environment by manufacturing particularly tough plasma membranes. These are

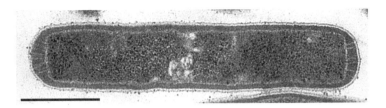

Figure 4.2 Electron micrograph of *Methanothermus fervidus.* In this and subsequent figures, the scale bar represents one micrometer. This methanogen illustrates a typical bacillus shape. (Micrograph courtesy of Professor Karl O. Stetter.)

based on particularly stable lipopolysaccharides: membrane-spanning isoprenoid lipids coupled to polymeric sugars, containing glucose and mannose.

We will discover that *Thermoplasma* is enigmatic in a number of ways. One is its preference for a coal mine habitat:clearly, hot mine wastes are a recent development in the long history of the Earth (and of life on Earth). It is thus difficult to imagine that these organisms evolved in the context of mining, yet they have turned out to be extremely difficult to find anywhere else. In particular, the sorts of *Thermoplasma* found near coal mines have proved noticeably absent in hot springs that might

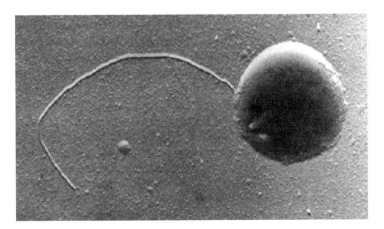

Figure 4.3 Electron micrograph of *Metalosphaera sedula.* A coccus with a single flagellum. (Courtesy K. O. Stetter.)

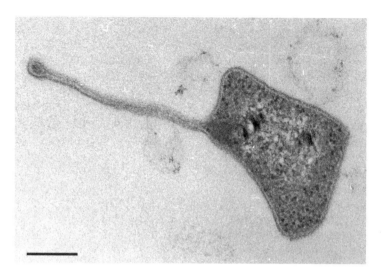

Figure 4.4 Electron micrograph of *Archaeoglobus lithotrophicus*. Note the irregular shape. (Courtesy K. O. Stetter.)

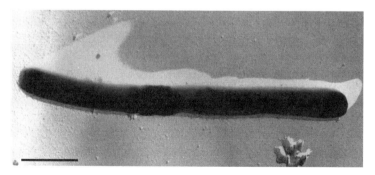

Figure 4.5 Electron micrograph of *Methanopyrus kandleri*. In this case, the cell was shadowed with a film of platinum. (Courtesy K. O. Stetter.)

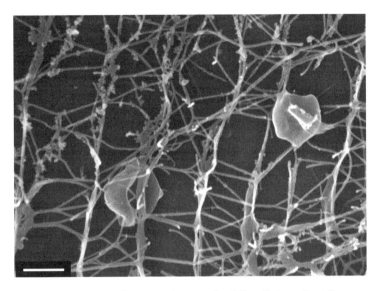

Figure 4.6 Scanning electron micrograph of *Pyrodictium abyssi* showing cells and their connecting network. (Courtesy K. O. Stetter.)

have seemed likely candidates (although they do have some close relatives in such places).

In contrast, the record for cells that are bizarre in their geometric precision must go to certain members of the halophilic branch of the *Archaea*. While many halophiles do appear as unremarkable rods and spheres, a number don't. The genus *Haloarcula* includes species with cells that are quite perfect rectangles or squares, looking, in the microscope, more like crystals than cells. The generic name, *Haloarcula*, appropriately enough, means salt box. And some of these square cells remain attached after cell division so that, in the end, they give the appearance of a sheet of postage stamps while others have been described as ribbon-like.

The first account of these wonderful square cells was first published by Walsby in 1980 and the impact of the discovery was such that the organism is still sometimes called Walsby's square bacterium. He had isolated the creature from extremely saline pools near Nabq, on the Sinai Peninsula. It has subsequently been identified in similar habitats in Mexico, Spain, and the Crimea. For some reason, the organism has proved

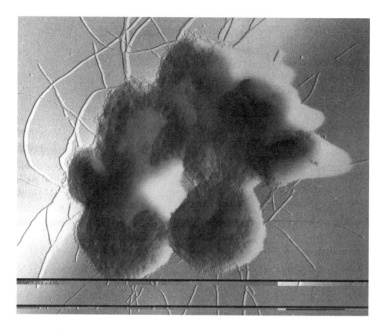

Figure 4.7 Platinum-shadowed electron micrograph of *Thermosphaera aggregans*. This organism receives its specific name from its tendency to form clusters, somewhat resembling a bunch of grapes. (Courtesy K. O. Stetter.)

extremely difficult to grow in culture, and all studies of it have necessarily been carried out using cells collected from natural populations. One imagines that its unusual cellular shape must originate in the geometry of subunits of its cell membrane or wall, and its lipid composition does differ from that of other halophiles. That is one reason to place it in a genus of its own.

Walsby's *Haloarcula* is found floating in extremely high numbers near the surfaces of pools where it lives, obtaining flotation from the presence of numerous gas vacuoles. Because the cells are so thin (approximately 0.1 micrometer) and so transparent, the vacuoles are often the only noticeable features seen in the light microscope. But the cells are also often noticeable in being motile: they are provided with flagella located at their corners and can move about in a highly coordinated fashion.

Cells of still other halophiles occur as triangles. Cells of *Haloarcula japonica*, first isolated in 1986 by K. Horikoshi from Japanese salt ponds, are perhaps the most dramatic. Often, they are in the form of a perfect equilateral triangle, although they can be merely isosceles, rhomboidal, or even square. The cells are motile, moving by means of tufts of flagella that are situated strategically on one apex of the triangle. The occurrence of triangular cells leads, at once, to the question of cell division. Most cells in nature exhibit bilateral symmetry—they are based on a plane of symmetry such that the two halves are mirror images. It is inherent in the geometry of such cells to divide into two more or less identical offspring, the products of the symmetrical cleavage. But how should a triangular cell behave? It exhibits bilateral, but also radial, symmetry, so that it is difficult to know a priori how division should proceed.

In fact, when cells of *Haloarcula japonica* divide, there are two, and not three, "daughter cells." Division occurs either on the plane formed by an apex and the opposite midpoint, or that between two midpoints. For this reason, the products of division can be either triangular or square, and square cells are apparently able to subsequently make the transition, via a rhomboidal intermediate state, to a triangle. This mode of cell division accounts for the range of cell shapes (square, rhomboidal, and triangular) that are in fact observed in cultures of these creatures.

What can we conclude from this puckish variety in the shapes of archaeal cells? Little, perhaps, beyond noting that these organisms must have some unusual ways of manufacturing cell-wall components and assembling them to form walls. But it would appear that these *Archaea* may have evolved relatively little from ancestors that had not yet become fully committed to specific ways of making cell walls or, perhaps, even of dividing. Thus, in the impressive variety of archaeal cell shapes, we may be observing vestiges of an era when experimentation with such matters was still in process.

Where They Live
(and How They Manage)

Where do *Archaea* live? The short answer is "everywhere." It has become clear from rRNA analysis that representatives of the *Archaea* occur widely throughout the biosphere. But there are certain locations where they are especially likely to be encountered. These places tend to be somewhat remote from the ecological mainstream: relatively obscure nooks and crannies in which rigorous environmental conditions have sheltered the *Archaea*, apparently for very long periods of geological history. These locations have historically provided inocula from which microbiologists have isolated pure archaeal cultures. And they are also locations that raise an obvious question: how do thermophilic, acidophilic, or halophilic *Archaea* manage to live in such uncompromising environments? In other words, what physiological adaptations have evolved to make such "extremophile" life possible?

Where Methanogens Live

The first rule of methanogens is that they don't tolerate oxygen. Therefore, they are only found in anaerobic environments, either environments rich in other microorganisms that consume oxygen, or where nonbiological chemical processes do the same thing. Take, for example, anaerobic muds, known since the time of Alessandro Volta at the end of the eighteenth century to be sites of methane production. These include bottom muds of swamps and marshes, those from the beds of fresh and marine bodies of water, as well as muds originating from human activities: sewage plants and rice paddies. Paddy fields are particularly good habitats for methanogens and account for a major part of worldwide methane production—more than all natural swamps and marshes combined. Indeed, even though marine methanogenic *Archaea* are widely distributed, marine environments are relatively low in methanogenic activity when compared to freshwater ones.

In some instances, methanogens live in small anaerobic pockets in an otherwise oxygen-rich area. Such regions are often formed by the action of microorganisms that locally consume all the available oxygen. These pockets evidently occur even in the open ocean, accounting for the small amounts of methane that can sometimes be found there. If suitable levels of moisture are available, they also occur in porous stone, even within the masonry of ancient buildings (notably, in Germany and Italy). There, brick, limestone, and sandstone harbor local partnerships (consortia) of microorganisms that include archaeal methanogens as well as bacterial methanotrophs, organisms that obtain energy by oxidizing the methane produced by their partners.

The Song of the Camel

It is said that groups of camels are audible over great distances, and that in desert wars, camels are not to be considered vehicles of stealth. It is not that these "ships of the desert" engage in idle chatter: their characteristic sound, rumbling over the sands, is the explosive noise of eructations or, to use the cruder word, belches. Camels, like cattle, goats, and sheep, are ruminants and, as such, are producers of methane; the release of the gas is often a noisy process. And such ruminants produce methane to the extent that they overshadow even the global contributions of paddy fields and swamps. Indeed, increasing worldwide populations of domestic ruminants may account for much of the increased release of the greenhouse gas methane to the atmosphere during recent decades.

Ruminants are distinguished by an organ, the rumen, in which a mixed population of microorganisms, including *Eubacteria*, methanogenic *Archaea*, and anaerobic (eukaryotic) protozoa, carries out digestion of cellulose and other polymeric sugars. The rumen is a cul de sac in the digestive system into which chewed plant material passes. There, microorganisms hydrolyze the polymers to their sugar subunits. Others transform these sugars into acetic and other acids, which are absorbed into the animal's bloodstream and, after being oxidized in tissue mitochondria, used to provide energy. Still other bacteria convert sugars into vitamins and amino acids, which are likewise absorbed by the animal's digestive tract.

Some of these transformations are oxidations, using up any free oxygen in the rumen and producing carbon dioxide. By keeping the oxygen concentration low, these reactions make it possible for methanogens to survive there and the carbon dioxide provides the starting material for the synthesis of methane. The plant material remains in the rumen for a number of hours, after which it is returned to the mouth for additional processing—chewing the cud. Finally, it enters a sequence of stomach-like chambers and then the small intestine, where much of the nutrient absorption into the bloodstream takes place.

A good deal of the carbon of the animal's food ends up by being converted into bacterial mass, so that the cud is rich in this material. In this fashion, plant material, relatively enriched in carbohydrate and poor in protein, is converted into protein-rich bacteria, which are subsequently digested and absorbed. These, in a sense, become the animal's real food, making such beasts efficient consumers of grasses and other austere provender. And, judging from the amount of methane produced in rumens, archaeal methanogens such as *Methanobrevibacter* and *Methanomicrobium* are major contributors to the conversion of carbohydrate into bacterial, and ultimately animal, protein. Consequently, these *Archaea* are major elements in this interesting symbiosis, in which the animal is supplied with food and the resident microorganisms, at least temporarily, with a good home.

Methanogens are also found in digestive systems of animals other than ruminants (e.g., in the stomachs of fish and baleen whales); the key requirement is always that the location is anaerobic. Even the human intestine is not innocent of the presence of methanogenic *Archaea* as evinced by the experiments of naughty children involving the (mild) detonation, by a lighted match, of flatulence. (The related practice of igniting marsh gas has provided similar amusement to generations of rural youth.) Thus, inventive children enjoy a place, albeit an obscure one, in the history of our knowledge of the *Archaea*.

Finally, one must mention the occurrence of methanogens in quite a different sort of intestine, and under quite different circumstances. Termites are able to digest cellulose owing to the metabolic activity of symbiotic protists* residing in their gut. Some of these protist cells, in turn, contain symbiotic methanogens resembling the free-living *Methanobacterium*. Such a cell-within-cell situation is termed endosymbiosis, which is extremely widespread in the microscopic world. In this instance, the methanogens use hydrogen and carbon dioxide from cellulose breakdown as the raw materials for methane production. Perhaps this is a case of adding insult to injury: not only do termites, with the help of their microbial symbionts, break down the cellulose of our (wooden) houses, but then they have the temerity to convert it to the greenhouse gas methane. In fact, termite gut methanogens are an astonishingly large source of atmospheric methane. The major biological contributors to global methane are the methanogens of natural wetlands, termites, rice paddies, and ruminants (according to some estimates, in that order). Moreover, any one of these sources is probably a greater producer of the gas than all industrial sources combined. It would clearly be a mistake to underestimate the global-scale ecological impact of *Archaea*.

In a similar fashion, some free-living protozoans, including ciliates, and amoebas that reside in aquatic sediments, also contain methanogenic endosymbionts. These are responsible for the large quantities of methane that we just learned are commonly produced in such places. Finally, methanogens are resident as symbionts in various eukaryotic organisms that lack digestive tracts and so don't harbor anything like a rumen flora. Some are, for instance, found in the heartwood of trees and others in the tissues of marine invertebrates.

---

*Protists are eukaryotic microorganisms. Many were formerly called protozoa, which denoted single-celled animals. We no longer consider them animals.

Where Halophiles Live

Strongly halophilic organisms require sodium chloride at concentrations several times greater than that of ordinary seawater. We observed that natural waters containing salt at such high concentrations occur chiefly in hot, dry regions of the Earth where evaporation is high and rainfall low. There are some exceptions, such as cold saline ponds on the antarctic continent, but these are rare.

The record in salinity is probably held by the Dead Sea, which also enjoys a second distinction, that of being the lowest body of water on earth—400 meters below sea level. The Dead Sea is situated in a geological rift valley that is both uncommonly dry and hot. It receives water from the biblical Jordan River and, having no outlet, evaporates most of it. In the Dead Sea, the sodium concentration is only about four times as high as in the ocean, but other minerals, particularly magnesium and calcium, are spectacularly elevated, and one observes encrusting deposits of their salts along the shoreline. The Dead Sea is, of course, not dead at all: it is home to the eukaryotic alga, *Dunaliella*, as well as populations of several halophilic microorganisms, chiefly *Archaea*. Population densities of the latter are sufficient to give the sea a distinctly purple color.

Evaporation pans, or salterns, have been employed in the Mediterranean basin for the production of sea salts since classical times. And there are Latin accounts of strange reddish discoloration of these salterns, with such staining undoubtedly leading to considerable anxiety among those engaged in the salt enterprise. Eventually, in the nineteenth century, the microorganisms that turned out to be responsible for the red "blooms" were isolated from foods that had been (imperfectly) preserved with sea salts from such locations. These were, of course, halophilic *Archaea*, although they were not recognized as such until a century or so later.

The environmental pH is an important factor determining which halophiles can grow in a particular place. Thus, of six recognized genera of halophilic *Archaea*, two have representatives occurring only in soda lakes, at a high pH. These are *Natronobacterium* and *Natronococcus*, organisms whose names serve to remind us that natron was the name for sodium carbonate in pharaonic Egypt. The four genera of halophilic *Ar-*

chaea that grow at a neutral pH are *Halobacterium, Haloferax, Haloarcula,* and *Halococcus.* One such species is *Halobacterium sodomense,* which was isolated from the Dead Sea, in the vicinity of the site of biblical Sodom. The Dead Sea is noted for its high magnesium content; not surprisingly, this organism requires high magnesium for growth.

Where Thermophiles Live

Practically all the contemporary biosphere is cool: locales exhibiting a temperature of more than 50 degrees are quite unusual, and many organisms live their entire lives below 20 degrees. Hot environments are restricted to geologically active locations, where heat escapes from deep in the Earth, often transported in the form of geothermally heated water. These regions, in turn, are usually found along the boundaries of geological plates. In such places, either new crust is being formed by upward extrusion of magma with subsequent spreading of the cooled product, or plates are colliding, one riding on top of the other, with crust material being returned to the Earth's depths.

Such geologically active regions, often studded with volcanoes and prone to seismic movement, also contain heated springs. These result from a circuit of water movement: surface water percolates downward until it comes into contact with heated rocks. The rocks are, in turn, heated by magma from still deeper layers. From the rocks, the water receives heat and dissolved minerals and is then driven upward by steam and its own reduced density, due to heating.

Magma, arising from 100-kilometer depths, and composed of molten rock, has a temperature in excess of 1000 degrees. Of course, the water that it ultimately heats never achieves such a temperature, as nonmolten rock is usually interposed. And, if the rising water is diluted by enough cool seepage from the surface, the final result is a modestly warm spring, with a temperature ranging from ambient to, perhaps, 50 degrees. If less dilution occurs, the water may emerge at its boiling point— 100 degrees, if it is a terrestrial spring, and as much as 400 degrees, if it is a hydrothermal vent on the seafloor and under elevated pressure.

The microbiological study of hot springs has been worldwide, including exploration of springs in the American West, Japan, Iceland, New Zealand, and Africa. The springs are exceedingly variable in their chemistry, temperatures, and plumbing details, such as flow rate and periodicity (e.g., geysers). In view of such variability, it is significant that some thermophiles, such as *Sulfolobus,* appear to be widely distributed, found virtually everywhere they have been sought.

A particular variation on the hot-spring motif that has harbored many thermophiles, and given its name to at least one, is the solfatara. The name is an Italian place name, that of a site not far from Mt. Vesuvius. A solfatara is a rather diffuse hot-spring area, where hot gas, including hydrogen sulfide and steam, heat the soil and provide homes for amenable microorganisms. Deeper strata of such soil favors anaerobic organisms, shallower levels, oxygen-consuming microbes. In general, solfatara fields are characteristic of well-established, even declining, hydrothermal activity and it is likely that the Italian examples were extant in the time of Pliny the Elder (who died nearby during a volcanic eruption in A.D. 79).

As the seabed has become progressively well explored, marine hydrothermal vents have turned out to be common. This is because the plate boundaries are often located in ocean basins, where the crust is particularly thin, the continents being regions of plate that are thickened by an overlay of less dense material. (In other words, the ocean beds are "pure" plate, without the overlay and, hence thinner.) And, new plate material—mostly basalt—is formed by magma upwelling in geologically active regions of the mid-ocean, where the vents occur. Spreading in both directions from these regions, the new crust pushes the continents apart at a speed of around one centimeter per year.

One such hydrothermal region is the mid-Atlantic ridge, roughly halfway across the Atlantic Ocean. The ridge surfaces briefly in Iceland, where terrestrial hot springs abound and vulcanism is vividly apparent. Because new seafloor is constantly produced along the ridge, the floor is much older at the ocean boundaries than at its center, and the spreading process continuously contributes new crust to the Earth's surface.

Although the first submarine vents were not actually observed until the 1970s, there were several clues that earlier alerted scientists to their possible occurrence. For one thing, in the 1960s sediments dredged up from the vicinities of mid-oceanic ridges were found to contain iron and manganese oxides. It is proposed that these originated in the interaction of hot crust material, which was rich in iron and manganese, with cold seawater. Chemical analysis of seawater and, in particular, measurement of the relative amounts of isotopes of helium, also led to the conclusion that hydrothermal processes made important contributions to seawater chemistry and, through precipitation, to marine sediments. It is not an exaggeration to say that chemical inputs from marine springs are at least as important in determining the chemical composition of the oceans as are the contributions of all the river-borne products of continental weathering, long thought to be paramount. Therefore, no one was much surprised when numerous active marine vents were observed along a number of widely distributed regions of seafloor spreading. But it was their biology that led to plenty of surprises.

The Dark World of Black Smokers

Oceanographers are fond of pointing out how little we have known, until very recently, about the deep sea bottom—although it comprises a significant fraction of the total surface of the planet. There was a time in the 1970s, they tell us, when we were more familiar with the surface of the moon, having been there, than with the deeper parts of the ocean. But about when some humans were first treading on the moon, others were making the first, tentative voyages in a new breed of submarines: the research "submersibles" able to withstand the pressures found at great depths.

When these first "live" explorations of mid-ocean ridges were made, there were already chemical and isotopic reasons to expect something interesting to turn up. Remote temperature measurements of seawater had disclosed "temperature anomalies"—abrupt thermal discontinuities that one suspected were due to hot water emerging from hypothesized vents. Water samples obtained at such temperature anomalies also exhibited helium isotope ratios suggestive of hydrothermal activity. So it came as no surprise when scientists on the sub-

Figure 5.1 The deep submersible *Alvin* emerging from a dive. Newly acquired samples are in the foreground. (Courtesy K. O. Stetter.)

mersible *Alvin*, which hailed from the Woods Hole Oceanographic Institution, observed the first hydrothermal vents. These were found at a depth of about two-and-a-half kilometers on a part of the Galapagos ridge. The hot vent water, less dense than the cold seawater around it, shimmered like air over a hot road. But, as marvelous as it was to see the vents, it was even better to discover that they served as a sort of biological oasis in the otherwise barren deep ocean. Surrounding the vents were shoals of giant clams, mussels, crabs, pink fish, and purple sea anemones. This seemed most odd, as the deep ocean, below levels where there is sufficient light for photosynthesis, usually supports little life and, especially, few organisms of any size. Surprisingly, it would turn out that uncommonly large invertebrate organisms are a hallmark of vent-based communities.

Vent Communities

Other geothermal localities turn out to be enlivened by the presence of giant (two-meter) worms belonging to the animal phylum *Pogonophora*. These worms completely lack a digestive system, an omission obviously raising important nutritional issues. Indeed, this entire, flourishing vent community raised an obvious nutritional question: how do these large and locally numerous organisms obtain nutrients in the relative desert of the deep sea? The answer clearly must have something to do with their proximity to the vents and vent water. In fact, it has everything to do with it.

Vent water, it turned out, is loaded with hydrogen sulfide, H_2S. When it mixes with seawater, it cools rapidly from as much as 350 to less than 5 degrees, and various sulfides precipitate. In many instances, black particulate iron sulfide gives the vent water the appearance of black smoke—hence, the term "black smoker." Often, insoluble metallic sulfide—iron, copper, and zinc—precipitate and form tall chimneys through which the vent water emerges. In addition, dissolved calcium ion in the hot vent fluid meets sulfate from the seawater, forming calcium sulfate and augmenting chimney construction.

It seems that a great deal of sulfide chemistry is going on around the vents, and sulfide is also at the heart of the nutritional question. Thus, the immediate surroundings of vents contain populations of *Eubacteria* able to use oxygen dissolved in the seawater to oxidize hydrogen sulfide. These include species of the genera *Thiobacillus, Thiothrix,* and *Beggiatoa,* the last often occurring in the form of thick mats. But sulfides are not the only components of biological interest in the vent efflux: there are often also considerable quantities of dissolved manganous ion, molecular hydrogen, carbon monoxide, and sometimes, ammonium ion. And the bulk seawater, into which the vents pour, contains dissolved oxygen, carbon dioxide, and bicarbonate. All these provide raw materials for a complicated web of life processes and contribute to the high productivity of the vent ecosystem.

Water from submarine vents often resembles water obtained from terrestrial hydrothermal springs with regard to chemical composition. After all, the geological processes underlying both have a great deal in common—both are frequently part of the same plate-spreading systems, as where the mid-Atlantic ridge rises above the surface in Iceland. But one common difference between marine and terrestrial hot springs deserves mention: terrestrial springs are often acidic, submarine springs, seldom so. More exactly, it appears that terrestrial springs fall

Figure 5.2 A terrestrial hot spring from near Hveravellir, Iceland. (Courtesy K. O. Stetter.)

into two classes: acidic springs, with a pH of around two, due to large amounts of sulfuric acid, and neutral springs, with a pH of around eight, due chiefly to the presence of bicarbonates. As a rule, geysers usually spout neutral water, while slowly seeping springs tend to be of the acidic sort. And neutral springs are usually long-lived, persisting sometimes for centuries, whereas acidic springs often expire after only a few years. In any case, submarine vents are usually of the neutral variety and most thermophilic *Archaea* isolated there are not acidophilic. In contrast, most *Archaea* that are both acid and heat lovers, like *Sulfolobus*, are from springs located on land.

Vent Productivity

In most of the living world, photosynthesis is the entry point for carbon and energy, with the carbon being "fixed" (transformed into organic carbon) from carbon dioxide. In other words, photosynthetic organisms are engaged in "primary production." Primary, in this context, means being the first step in a food chain. In the vent communities, the carbon also comes from carbon dioxide, but the energy, rather than being provided by sunlight, is associated with the oxidation of sulfide by the bacteria mentioned above. Energy for primary production in these systems is thus from mineral oxidation, and the organisms responsible are termed lithotrophic. Of course, different vent systems have different chemical properties and different sorts of bacteria must play a role. For example, some vents contain methane, and are colonized by methane-oxidizing *Eubacteria* that obtain their energy from that reaction. In contrast, other vents are sites of methane production, with methanogenic *Archaea* using hydrogen to reduce carbon dioxide.

Free-living bacteria of the sorts that we have been describing may play a less significant role in vent communities than do bacteria living in symbiotic association with certain eukaryotic organisms. For example, the giant *Pogonophora* worms that lack digestive systems do have specialized organs—trophosomes— that contain prokaryotic cells. These cells turn out to be sulfur-oxidizing *Eubacteria*, closely similar to the free-living *Thiovulvum*. The cells also are able to utilize carbon in the form of carbon dioxide and so are appropriate initial participants in food chains. The giant worms, it appears, obtain their nutrients in the form of waste products and dead cells from their resident bacterial population. And the other "giant" organisms of the vent community, mussels and clams, have similar bacterial populations located in their gills. Each large organism has its own specific bacterium, which presumably evolved in parallel with its host.

But free-living microorganisms are also present and can oxidize various chemical forms of sulfur and nitrogen as well as iron, manganese, hydrogen, and methane. Many of these are *Eubacteria*, but many are thermophilic *Archaea* as well, the latter closely associated with the oxidation of sulfur.

Water can issue from black smokers at temperatures well in excess of 300 degrees. (High pressure makes this possible. In fact water at a depth of 2600 meters boils at about 450 degrees.) This water is surely sterile, as 300 degrees is well beyond the limit imposed by the instability of crucial molecules like ATP at higher temperatures. There have been reports of microorganisms being isolated from water at around 150 degrees; these have not been confirmed and are almost undoubtedly

Figure 5.3 Black smoker chimney from the East Pacific Rise at a depth of 2500 meters. (Courtesy K. O. Stetter.)

artifacts, resulting from contamination. However, such superheated hydrothermal water mixes rapidly with seawater in and around the chimneys and other vent structures, so that a temperature gradient occurs. The same mixing process leads to chemical gradients as well, as the various vent-water constituents mix with the bulk ocean water. These gradients provide ample opportunities for thermophilic organisms to live and it is there that many of the thermophilic *Archaea* have been found. Keep in mind that the advantage of living in such a chemical gradient, resulting from vent efflux, is the way in which nutrient materials are continuously being supplied and waste products continuously washed away. A charmed life, one might think.

One would be better able to judge how "charmed" archaeal life really is if one had a better notion of their precise habitats. In particular, thermophilic *Archaea* reside in hydrothermal systems, but we often don't know exactly where, nor are we very certain about the dimensions of population distributions. On the one hand, they may be visitors from very deep parts of the system, perhaps from chambers quite deep in the Earth's crust. On the other hand, they may only reside in the terminal parts of the system, in sediments close to active vents, or even in nearby seawater that receives, and mixes with, hot vent effluent. Investigation of thermophile distribution in geothermally active locations suggests that these organisms may reside in extensive regions of the seafloor, responding to heat from interaction between deep magma and the shallower crust. When new vents or areas of seepage are established, the organisms are released to the sea, becoming accessible to study.

Surviving (and Flourishing) in Extreme Environments

We know through rRNA sequences that the *Archaea* and *Eubacteria* branched a very long time ago and that at least their thermophilic representatives evolved slowly. Such slow evolution appears associated with residence in harsh environments that blunt some effects of natural selection. Organisms that live in boiling springs are not troubled by competition from creatures for whom high temperatures are fatal. Moreover, there is reason to believe that such springs are permanent features of the earth over great intervals of geological time. Thus, one of the usual driving forces for organic evolution, an evolving environment, is diminished.

Likewise, few organisms can stand saturated salt solutions, with their dehydrating propensity. Therefore, extreme halophiles, whose defenses against concentrated salts will be discussed in Chapter 7, also experience little competition, and their rate of evolution is likewise diminished. This observation leads to two questions. First, since life at 100 degrees, or in a concentrated salt solution, is clearly possible, why haven't other organisms evolved in order to exploit such an environment? The answer, like the answer to many biological questions taking a similar form, may simply be that they just didn't. Natural selection cannot be assumed to have encompassed every option; indeed, there may not have been sufficient time. After all, the four billion years, give or take a little, that have been available for evolution is really a rather short time when one considers the number of possible biological permutations.

The second question is: how do they manage to do it? Enzymes, as a rule, are destroyed by even short periods above 50 degrees, whereas some of the organisms in questions live at 100 degrees. (Consider a boiled egg: the egg white, like enzymes, is protein, and is clearly heat-sensitive.) Also, membranes become intolerably fluid at these elevated temperatures and the double helix of DNA "melts" into single strands at temperatures ranging from about 80 to 100 degrees, depending on base composition (i.e., the GC ratio).

Indeed, the connection between DNA stability and base composition leads to something of an enigma: a high guanine plus cytosine content—that is, a high GC ratio—promotes stability in DNA helices. However, many hyperthermophiles are characterized by a low ratio, just the opposite of what one might expect. For example, the terrestrial thermophile, *Acidianus*, with a growth optimum of about 90 degrees exhibits a record GC ratio of 31 percent and virtually the same numbers apply to the marine-vent archaeon *Methanococcus*. It is evident that factors other than base composition must account for the DNA stability (and indeed, survival) of these organisms.

Even small molecules like ATP (the energy carrier) and NAD (an oxidation coenzyme) become unstable at high temperatures. Both of these exhibit a half-life of about thirty minutes at 100 degrees; at the very least, they would need to be replaced in an aggressive way. And yet, thermophiles such as *Pyrodictium*, denizens of deep-sea vents, where high pressure elevates the boiling point of water, grow at 110 degrees! How do they protect their proteins and other macromolecules against thermal destruction?

Similar questions need to be considered with respect to halophilic *Archaea*. High salt concentrations can also destroy protein structure, harm membranes, and, especially, dehydrate cells. Most organisms are killed by saturated salt solutions; how do the halophiles survive? Only those methanogens that are not also thermophiles are immune to such stability problems. Their particular form of "extreme environment" is an anaerobic one and the absence of oxygen is definitely not destabilizing to the molecules required for life. Indeed, the *presence* of oxygen is more likely to lead to instability—for example, through the formation of oxygen free radicals.

Protection Against Excessive Heat: The Protein Problem

First, consider the hard-boiled egg. The white of an egg, prior to boiling, is a concentrated solution of protein, containing one major constituent called egg albumin. This protein is stable for long periods at room temperature, which is why unbroken eggs do not require refrigeration. But, if the temperature is raised to, say, 70 degrees, gradually the egg white becomes opaque and solid. Biochemists say that the egg-white protein has become denatured; the rest of us would call it cooked.

Here is what happens on the molecular level. Albumin is a globular protein, which means that its *overall* shape is roughly ellipsoid. But proteins are actually chains of amino acids, so we should picture this globular protein as a strand, wound in space in a particular way and enclosed within the boundary of the ellipse. The individual amino acids are linked in the chain by strong (covalent) bonds; the bonds that hold the chain in its correct three-dimensional structure are weaker and mostly hydrogen bonds.

Covalent bonds result from sharing electrons between atoms. The much weaker hydrogen bonds are electrostatic forces between relatively negatively charged atoms, like oxygen and nitrogen, and the relatively positive hydrogen. Hydrogen bonds are also responsible for the helical structure of DNA and the solidity of ice.

The three-dimensional placement of the amino acid chain is also the result of electrostatic forces between charged amino acids, on the familiar basis that like charges repel and unlike charges attract one another. The charges on amino acids change as the pH of the protein's environment is altered—this is why extremes of pH also alter protein structure, often in a destructive fashion.

Undamaged protein has a unique three-dimensional structure and a unique pattern of internal (intramolecular) hydrogen bonds. But when it is heated, hydrogen bonds are broken and new, one might say "incorrect," ones are formed. Indeed, when the heating process continues long enough, hydrogen bonds start to be formed between different molecules—intermolecular bonds. Then the protein becomes insoluble, achieving the state observed in a boiled egg. Notice that this process can also be accomplished by an extreme pH via alterations in the charges on amino acids: the effect of boiling an egg can be imitated by treating the egg with acidic lemon juice or vinegar. (In Latin American cuisine, fish may be "cooked" in this fashion with lime juice; the result is ceviche.) And, whether using heat or acid, if the process goes far enough, enough intermolecular bonds are established to ensure irreversibility. Or, in other words, it is exceedingly difficult to unboil an egg.

The thermal stability of a protein is determined by the numbers and strength of these hydrogen bonds, and other electrostatic forces, that hold the protein in its correct three-dimensional structure. Alteration of that structure alters stability. And it is evident that the three-dimensional structure of a protein is determined explicitly by the sequence of amino acids in the protein: a particular structure reflects a particular sequence, and vice versa. So, the causes of protein stability and instability must be sought in the amino acid sequences (and only there).

Amino acids influence protein stability in another way. A stable protein tends to have nonpolar amino acids in its interior and polar amino acids on the outside—where water is usually located. In other words, the most probable and therefore most stable situation is the one in which nonpolar amino acids are protected from having to interact with (polar) water. Note the similarity with the arrangement in membranes, where the nonpolar fatty acids are protected from water—again, the most stable configuration.

Proteins from thermophiles are stable at higher temperatures than those from ordinary organisms. ("Ordinary" here means growing with an optimum temperature in the range of 20 to 40 degrees. Such organisms are called mesophiles.) Moreover, if a protein has a measurable function, such as enzymic activity, then this activity is likely to be maximum at a higher temperature in the case of a thermophile. Indeed, the temperature optima for enzymes tend to correspond quite closely to the optimum growth temperature of the organism.

In trying to discover the basis for protein thermostability, an obvious approach is to isolate a comparable enzyme from a mesophile and a thermophile, and see how their amino acid compositions differ. When this was first done, the expectation of just about everyone was that thermostable proteins would turn out to have grossly different sequences than those of corresponding mesophilic ones. In fact, however, thermophilic proteins were astonishingly similar to corresponding mesophilic ones. Often there are extensive regions of homology between the two, regions in which the amino acids line up perfectly. But one observation seems to shed light on the matter:

proteins from thermophilic cells do usually have significantly more nonpolar amino acids. For example, in the extreme thermophile, *Pyrococcus furiosus*, the ratio of nonpolar to polar amino acids in a particular enzyme, enolase, was around three, whereas in a number of nonarchaeal, nonthermophilic organisms, the ratio was only around one. Presumably, stability is associated with the tendency to internalize those nonpolar amino acids molecules, although we don't know exactly how the whole thing works. The authors of the study in question thought that the distinction between the (nonpolar) inside and (polar) outside of the enzyme was more "distinct" in the thermophilic case.

Heat Shock Proteins

There is a second way in which proteins achieve heat stability. Many organisms respond to an elevated temperature by synthesizing special "heat shock" proteins, also called chaperonins. These are heat-stable themselves, and bind to other proteins, protecting them against thermal damage. Usually this strategy can give cells about 5 degrees worth of protection and, for most organisms, is a short-term remedy—the chaperonins are only made as long as the heat stress persists. But many thermophiles make chaperonins on a permanent basis, and these may be a contributor to their long-term viability at high temperatures.

Synthesis of such protective proteins is not without metabolic cost. When grown at its maximum temperature of 110 degrees, the extremely thermophilic methanogen *Pyrodictium* manufactures a chaperonin to the extent of its composing about 80 percent of the cell's total protein content. Such synthesis requires an enormous outlay of ATP and other energy sources. In contrast, when grown at a mere 100 degrees, *Pyrodictium* makes very little, so that gaining that last 10 degrees of scope is very costly indeed.

There are also comparable proteins that appear to protect nucleic acids and, for example, can prevent the melting (strand separation) of double-stranded DNA. Such a mechanism may protect thermophile DNA from high temperature: several thermophilic *Archaea* have been observed to synthesize proteins that bind to their DNA, proteins that do not occur in organisms that live at lower temperatures.

Finally, proteins and other macromolecules can be protected against thermal damage by smaller bound molecules. For example, cells of *Methanopyrus fervidus*, one of the most extreme hyperthermophiles, able to grow at 110 degrees, contain high concentrations of a small compound related to glycerol—namely, cyclic 2,3-diphosphoglycerate. This appears to protect the proteins of this organism against thermal damage and it is noteworthy that the concentration of the compound in the cell is closely correlated with the temperature at which the organism is grown.

Protection from High Salinity

Life at high salt concentrations resembles life at high temperatures in the sense that proteins are unstable in either instance. In the case of the salt problem, there are two potential lines of defense: the plasma membrane and the stability of the proteins themselves. The plasma membrane of halophiles does not exclude salts, so that the overall internal salt concentration is high. However, it is able to exercise selectivity, so that the ratios of specific ions constituting the salt can be modified. In particular, in highly concentrated brines, there is usually a great deal of sodium, but much less potassium. However, many cell functions require potassium, and sodium often interferes with its action. Therefore, halophilic cells usually exclude sodium, while concentrating potassium, often ''pumping'' it into the cell against a thousand-fold concentration gradient. In this fashion, the total ionic strength remains the same on both sides of the plasma membrane, but potassium is the prevailing cation on the inside.

The second element of defense is the ability of the enzymes within the halophilic cell to tolerate high potassium. This tolerance is not observed with nonhalophilic organisms, whose enzymes are rapidly destroyed by high salt. Such destruction is effectively the same as heat denaturation. Indeed, enzymes from halophilic *Archaea* actually require high salt concentrations, generally in the form of high potassium, for normal function and stability. This altered sensitivity to salts is associated with subtle modifications of their amino acid composition when compared to those of organisms that live in lower salt environments.

Features of halophiles enabling them to tolerate concentrated brines also appear to produce a generally rugged constitution. Thus, early bacterial studies of the Dead Sea were carried out in 1936 by B. E. Volcani, who kept sealed samples of enrichment cultures for almost sixty years. Even after that length of time, a number of different halophilic *Archaea* remained viable, and it proved possible to isolate pure cultures of several of them. These were characterized using PCR and subsequent sequencing of rRNA genes, so that the techniques of the mid-1990s were applied to samples taken in the mid-1930s, an era when the region surrounding the Dead Sea must be presumed to have been a very different place. The isolated halophiles were identified as belonging to the genera *Haloferax, Halobacterium*, and *Haloarcula*, and may have included two new species of the last named. Patience is always a welcome attribute in a biologist, but an experiment lasting six decades must be something of a record.

From a Historical Perspective

What's In a Name?

The names *Archaea* and the earlier *Archaebacteria* were obviously intended to denote "archaic," both having been coined at a time when it was believed that these organisms might be of more ancient lineage than any others. Now that view has been modified somewhat, and we reckon that the archaeal and eubacterial lineages are of equal (and great) age, so that the most important landmark in early evolution becomes the point at which the two prokaryotic groups branched and became distinct. The current view is that this happened about 3.5 billion years ago. The thermophilic members of both groups do appear to be the most slowly evolving—the most ancient in these two very old domains. So especially among these thermophiles, eubacterial and archaeal, one may reasonably expect to find clues about early life and the environment that shaped it.

It is important to be clear about the meaning of "early life." The ancestral prokaryotic cells, about which our modern thermophiles may presumably offer clues, are themselves products of evolution. Living during the first of the roughly four billion years of life, they are hardly the first cells, about which we will probably never have very much reliable information. We should not expect our archaeal investigations to shed much light on first life, nor on the processes by which it arose. But if we succeed in learning something helpful about the time when *Eubacteria* and *Archaea* diverged, we will have accomplished a great deal, as that divergence formed the context for a great deal of subsequent evolution.

At Home on the Early Earth

It is likely that some properties of the *Archaea* reflect their environmental history on the Earth of three, or more, billion years ago. Some environments, in which present-day *Archaea* live, exhibit continuity with comparable environments on the surface of the young planet. For instance, hydrothermal springs, whether marine or terrestrial, are now quite sparsely distributed, being confined to regions of particularly great geological activity. But earlier hydrothermal systems were much more widespread because much more of the Earth's crust was geologically active when the planet was young. Therefore, such sites may be considered scattered vestiges of their earlier wide distribution. And the life that they support, such as thermophilic *Archaea* and *Eubacteria*, may perhaps be viewed as evolutionarily conservative remnants of early life that had formerly been much more widely distributed.

The notion of environmental continuity—of the persistence over long intervals of geological time of surface features, such as geothermal systems—requires an additional comment. In fact, a particular hot spring or submarine vent normally has a short life span, often a matter of only a few years. Thus, during the short time that submarine vents have been studied by scientists, whole cycles of birth and senescence have been recorded. But, as vents age and expire, others form and are colonized by organisms: it is the availability of these features that persists over immense spans of geological time. The reliable presence of vents, individually ephemeral but collectively unchanging, allows thermophiles to be so evolutionarily conservative. And vents tend to be clustered in geologically active regions, such as oceanic ridges and the environs of volcanos. Thus does proximity make them ripe for easy colonization and resulting population continuity.

The early Earth was hot, acidic, and sulfur-rich, so we should not be surprised that microorganisms that presently live in hot, acidic, and sulfur-laden environments appear to be the most direct descendants of early organisms. And the observation that many *Archaea* are anaerobic may reflect their having initially evolved in an atmosphere largely devoid of the gas—prior to the advent of oxygen-producing photosynthesis. In this fashion, living microorganisms may inform us about aspects of their history.

The Antiquity of Life

The history of the universe began somewhere between twelve and eighteen billion years ago, with the memorably named "Big Bang," followed by a long period of expansion and cooling. The solar system, including our Earth, formed something less than five billion years ago, probably condensing from the residue of collapse of first-generation stars. Then, by about four billion years ago, our Earth had solidified as a dense, spherical object, although it was still extremely hot (largely from gravitational contraction) and mostly molten. But at least some of the crust had solidified, as the earliest rocks found on the contemporary Earth are about four billion years old.

The earliest rock formations that contain recognizable microbial fossils are between 3.6 and 3.8 billion years old. These organisms mostly resemble rod-shaped bacteria, but there is enough variety to suggest that evolution had already been going on for a considerable time. Unfortunately, there is insufficient information to enable us to identify these cells as either *Eubacteria* or *Archaea* (or anything else). But it is striking that life has actually been present on Earth for most of its history and, indeed, the history of our solar system.

The Early Earth

Some Greenland rocks have been dated, using radioactive decay "clocks," at about 3.8 billion years. Several are volcanic components of the Isua formation and active volcanism was certainly a distinctive feature of the Earth of that era. In fact, the Earth's surface was magnificently unstable, a thin, mobile crust floating on a molten interior and punctuated with volcanic penetration from within. But certain other Isua rocks are sedimentary, having been formed in layers by means of sedimentation from an ocean. This is germane because it establishes that there was a significant marine environment around—that is, that there was water in liquid form readily available. The Earth, 3.8 billion years ago, was both hot and wet.

That initial heat was mostly the result of the compression of the Earth by gravity—the Earth had recently formed by accretion of bits and pieces in the solar system cloud and, in fact, was still in the process of formation. Indeed, it still is: the rain of meteorites that enliven our summer nights might be said to be the latest contributions of the primordial solar system cloud to the Earth. On (fortunately) rare occasions, larger objects, like asteroids, make their contributions with dramatic consequences.

Later, the Earth came to derive an increasing proportion of its internal heat from radioactive decay. The geological evidence of liquid water on the 3.8 billion-year-old Earth indicated that, by then, the Earth's surface had cooled to below 100 degrees, allowing water to condense as a liquid. Earlier, water certainly existed as steam in the atmosphere, but it is plain that the initiation of life depended on its condensation as liquid water. There is a great deal that we don't know about life's inception, but this much seems clear: it began in water, most likely in very hot (but liquid) water, and possibly in very hot water at an acidic pH. And it appears likely that those were the conditions that still obtained when the first *Archaea* and *Eubacteria* evolved.

The establishment of life in an environment of acidic water may indeed have set the tone for much that followed. Even today, all cells are full of water and bathed in a watery local milieu. And if acidic water (with its high hydrogen ion concentration) was the medium for early life, it is noteworthy that energy transfer in modern organisms is associated with hydrogen ion movement across membranes. (This is the case in both photosynthesis and cell respiration.) This evolutionary choice of hydrogen ion for energy transfer may thus reflect the high concentration of that ion that was present when (and where) life originated. Thus, a universal feature of cellular biology may be understood, at least in part, in terms of planetary history.

Perhaps the most important sense in which the early "pre-biotic" Earth differed from ours was the lack of geological processing due to the action of living organisms. It is clear that many of the chemical differences between the Earth and the other comparable planets, both in their surfaces and their atmosphere, can be accounted for by the presence of life here. For example, most of the oxygen in the modern atmosphere is due to photosynthetic life. On the other planets, where life has evidently not occurred, the atmospheres and geological components tend to be close to chemical equilibrium with one another, whereas, in our case, biological activity has displaced concentrations far from the equilibrium values. For example, the atmospheric oxygen concentration is much higher on the Earth than on, say, Venus, where it approximates the value expected from equilibria with oxygen-containing minerals occurring in the solid planet. Clearly, on Earth, oxygen-producing photosynthesis has left its mark. Even the oxidation states of iron ores can reflect the presence of life: there are banded iron mineral formations that represent successive layers of oxidative activity by microorganisms over long periods of geological time.

The Ocean

Rather little is known about the ocean of four billion years ago. One can guess from the composition of ancient rocks, and from known solubilities, that sodium was the prevalent cation (positively charged ion), as it is now. Likewise, chloride was the most common anion (negatively charged ion). It is also evident, from present carbonates in early rock formations, that carbonate from dissolved carbon dioxide was also. On the other hand, iron and phosphate—both significant for early life—were present at very low concentrations, but possibly higher in the vicinities of submarine hydrothermal vents. Vents may have represented important anomalies in other ways by providing locally elevated concentrations of biological raw materials.

Surprisingly, the ocean may well have been a quite transitory feature of the early Earth. It now seems likely that impacts of asteroid-sized objects transmitted sufficient energy to the planet to simply boil away the entire ocean so that it had to reconstitute itself by condensation from the atmosphere (as rain). In this fashion, ocean formation may have been a process that was repeated a number of times. We will return to this point.

The Atmosphere

The atmosphere of the young Earth has been the subject of considerable conjecture and our understanding of the topic has undergone a great deal of evolution during recent years. Much contention has centered on the question of how reducing the atmosphere was—that is, how many of its gaseous compounds were able to donate, as opposed to being able to receive, electrons. This question turns out to be of the greatest significance in deciding how atmospheric compounds might give rise to those small molecules, like amino acids and nucleotide bases, that are key ingredients of living cells. Also, the matter of how reducing the early atmosphere was is closely related to the exact way in which the Earth and its atmosphere was formed in the first place and, in particular, how rapid the process had been.

For a long time, there has been agreement that the oxygen content of the young Earth was extremely low, much lower than the one-fifth contribution to the total that it makes at present. It has also been accepted that major gases in the early atmosphere included methane (CH_4), carbon dioxide (CO_2), ammonia (NH_3), and molecular nitrogen (N_2). Formerly, it was believed that carbon monoxide (CO), molecular hydrogen (H_2), and various sulfides were also major components, but, more recently, these have been relegated to relatively minor status. As the members of this last group of gases are more reducing than the others, the trend in our thinking has been from a strongly reducing atmosphere to one that was much more mildly reducing. The importance of this conclusion will be apparent when we turn to the "prebiotic" chemistry that predated life itself.

Finally, another trace component of the ancient atmosphere was probably hydrogen cyanide (HCN), a gas that may have played a role in prebiotic chemistry as a synthetic reagent. It is somewhat amusing that two legendarily noxious chemicals, cyanide and carbon monoxide, were components in Earth's early atmosphere and were probably important in setting the stage for the origin of life. Now, both gases are considered to be extremely toxic, at least as far as oxygen-requiring organisms are concerned. This is because both are potent inhibitors of cell respiration (i.e., oxidations leading to energy production). Of course we breathers and users of oxygen may be forgiven a certain prejudice about this matter. A methanogen (if able to comment) might not share this prejudice.

Meteor (and Other) Impacts

The Earth, as well as the other planets and their satellites, was formed through the gravitational accretion of material from the primordial solar system disk. The disk was also the source of raw material for the sun itself. The disk material was, as we mentioned, probably debris from the collapse of an earlier generation star. Apparently, the formation of the Earth was initially quite rapid, with a sizable planet being collected together in approximately 10^7 years. But, long after the Earth had approached its full size, the rain of solid objects continued, but with the average size of the objects diminishing in time because the larger ones were the first to be gravitationally trapped. The rain continues even to the present day, but, fortunately, most of the objects, known as meteorites, are quite small usually in the millimeter range.

But, for the first half billion years or so, the impacting objects were large enough to transfer enormous energies to the Earth. In one such cataclysmic impact, the energy was apparently sufficient to expel a piece of crust that became our moon. But, probably more often, impacts were only of sufficient force to do major damage to the Earth, its atmosphere, and its oceans. For example, an object 100 kilometer in diameter, a size known from studies of the moon's craters to be a lively possibility at about 3.8 billion years ago, could transfer sufficient energy to convert roughly the top thirty-five meters of the entire ocean into steam. It is clear that if this sort of thing happened after life had originated, then life would have a much better prognosis for survival if it occurred in the deep ocean. Of course, land wouldn't provide any real protection at all, as the steam and other hot gases would have an excellent chance of eliminating any organisms residing there.

However, there were probably also objects hitting the earth that were considerably larger than 100 kilometers and these could have boiled the entire oceans away and, moreover, sterilized the entire Earth's surface with vaporized rock plus superheated steam at about 2000 degrees. Thus, it is entirely possible that life might have arisen and then been obliterated a fair number of times. And life might well have had quite different attributes each time. The life that we know clearly evolved from an ancestor that arose after the last such ocean-destroying impact event. And given the probable frequency of subsequent, if somewhat smaller, impacts, it seems likely that the earliest organisms that did survive to populate the Earth were to be found in the deep ocean, as far out of harm's way as possible. The vicinities of hydrothermal vents on the seafloor come to mind—and thermal vents were, as we saw, much more widely distributed then than at present. All of this brings us back to the *Archaea*.

When First Life Appeared

There is evidence from the dating of the moon's craters that, at about four billion years ago, major (i.e., ocean-boiling) impacts were still occurring too frequently for life to gain much of a foothold, but that by about 3.8 or 3.7 billion years ago the rate and intensity of impacts were falling off to the extent that organisms would be more likely to survive. By 3.8 billion years ago the Earth had also cooled to the extent of having a proper, (i.e., liquid) ocean, so, in at least those two respects, the stage was set. Therefore, it is significant that rocks dated at about 3.5 billion years contain microscopic objects that are undoubtedly cellular in nature. We should mention here that not all objects that look like cells really are. A number of alleged fossil cells have turned out to have nonbiological origin—to have been bubbles or some sort of mineral objects. Indeed, such apparent misidentifications should be kept in mind when claims are advanced for the occurrence of microbial fossils in Martian meteorites. But the fact remains that 3.5-billion-year-old rocks have been found to contain clear evidence of life, mostly, but not entirely, in the form of rod-shaped microorganisms.

But here is the most interesting aspect of all: some 3.5-billion-year-old rocks contain not only single cells, but also rather complicated structures called stromatolites, which are made up of a variety of prokaryotic cells, including some that form long filaments. These fossil stromatolites closely resemble living microbial mats that grow in shallow, tropical seas and some hot springs. Stromatolites, fossil or modern, contain a mix of organisms, including a variety of filamentous, photosynthetic bacteria. The observation of such diversity of prokaryotic microorganisms, and especially of complex stromatolites, indicates that by 3.5 billion years ago, considerable evolution had occurred, leading already to extensive biological diversity.

Of course, we don't know precisely when life first occurred, but it was probably between 3.8 and 3.5 billion years ago. Actually, available evidence places the time of origin closer to the former than the latter. This evidence includes somewhat contentious fossil observations, some isotope ratio measurements, and some estimates concerning the duration that would have been required to produce the observed prokaryotic diversity by 3.5 billion years ago. To understand the isotope ratio argument, consider that carbon occurs as several isotopes—forms of the element that differ in the number of neutrons in their nucleus. The isotope with an atomic mass of 12 is the most common, but there is also a significant amount of one with a mass of 13. Because chemical reactions associated with life favor the lighter isotope, the ratio of the two is different, when one compares carbon samples from a living or nonliving origin. There was a sharp transition in this ratio close to 3.8 billion years ago.

Thus, the gap between the time when life was possible, owing to availability of liquid water and the cessation of large meteor-impact events, and the time when it was clearly present may be, in geological terms, a very short one. This length of time required for life to appear (once the possibility existed) has profound implications for the question of what actually happened when first life emerged.

Some feel that life is highly improbable—that a large number of sequential events, all of low probability, were required for it to commence. Equally reasonable people argue, mostly on the basis of the same evidence, that life actually has a high probability—that life is virtually inevitable, once conditions become suitable. In other words, life must be considered a necessary feature of any part of the universe that has reached a certain stage of development. Clearly, these are the same people who are likely to advocate radio-frequency monitoring of remote parts of the universe—just in case "anyone" is broadcasting. In any event, it appears that emergence of the first life did not need to wait for billions of years once the Earth became sufficiently hospitable.

Where First Life Appeared

If complete certainty about the time of life's origin eludes us, so does the exact location of the glad event. Darwin argued for a "warm, shallow pond" on the grounds that useful organic molecules, the putative raw materials of life, could be concentrated there by evaporation. Besides, one imagines that the natural historian aspect of his character inclined him toward a habitat of known fecundity. Clearly, Darwin's shallow pond idea was a good one, and it has dominated thinking on the subject for well over a century, but other locations have recently been nominated as well.

For instance, atmospheric droplets have been suggested, partly because of the free availability of radiant energy and partly because the first cells could have plausibly originated from such droplets through acquisition (somehow) of a greasy coating. Another possibility is that the first life appeared on the interface between water and the surfaces of clays, iron pyrites, or other solid minerals. It has been pointed out that such surfaces could serve as binding sites, sequestering and concentrating chemicals, such as amino acids and nucleotides, that would be required for subsequent events leading to life. For instance, the binding of amino acids in close proximity to one another would favor condensation reactions, producing small polymers and, eventually, perhaps proteins. Moreover, clay surfaces have been shown to catalyze biologically relevant reactions such as the synthesis of nucleic acids.

The surfaces of iron pyrites carry out oxidation-reduction (i.e., electron transfer) reactions that might also have preceded life. Interestingly, certain thermophilic, sulfur-dependent *Archaea* have been observed growing on the surface of iron pyrites, which is, after all, an iron sulfide mineral. These organisms carry out the very oxidation-reduction reactions that have been suggested as playing a role in early life events.

A Hydrothermal Origin

It has been postulated that life first occurred in boiling hot springs, or vents, deep on the seabed. This proposal is clearly the antithesis of Darwin's hospitable pond. Arguments supporting this idea include the observation that the least evolved *Eubacteria* and *Archaea* are strongly thermophilic and that such hydrothermal systems are likely to be good refuges during hard times that might elsewhere accelerate evolution. (It is as if the microorganisms originated there and never left.) In addition, submarine vents can provide continuous flows of a promising mixture of nutrients, including some that are in distinctly short supply elsewhere.

For example, there is a certain mystery surrounding the importance of phosphate in information storage and energy transfer in living cells. Among other things, phosphate is a major component of the nucleic acids, DNA and RNA, and coenzymes and energy carriers such as NAD, ATP, and ADP. But phosphate is present at very low concentrations in the sea and in freshwater, so that it is usually the limiting nutrient for organisms that live there. Therefore, one may ask how it is possible that phosphate is so important to life while so rare in its environments. However, water in hydrothermal systems often percolates through phosphate-rich minerals such as basalt so that a rich phosphate supply might have been assured, and living things might have acquired a taste for it at the very outset.

Finally, an argument in favor of a hydrothermal location for early life rests on the observation that all cells are surrounded by membranes, which are composed of fat, or lipid, molecules. Lipids are able to form such a skin around cells due, in part, to their insolubility in water. It is hard to imagine a cell, even a primordial cell, that lacks a membrane, and there has been considerable discussion concerning the question of which came first in early life: membranes or nucleic acids? The answer is probably "both." This may be possible in the following fashion.

First, imagine a hot spring on the ancient seafloor. Heated water emerging from the vent would have already percolated through the rocks and clay of the seabed and would be charged with a variety of dissolved substances as well as suspended clay particles. These particles would transport bound molecules that are precursors of life, including amino acids and nucleotides, as well as small polymers formed from them: polypeptides and polynucleotides. In addition, there would probably be lipid molecules present, some of them likely being soluble in hot water, but insoluble in cold. Then, according to the scenario, the hot water would cool on mixing with seawater, and the lipid molecules would precipitate from solution, forming spherical vesicles. If some fraction of those vesicles contained some of the suspended clay particles inside, with their burden of bound amino acids, nucleotides, and so on, the stage might be really set. The raw materials for protein and nucleic acids would be internal, in close proximity to the clay surfaces, which could catalyze the formation of protein and nucleotide polymers. And a lipid membrane would tidily surround the whole business.

In fact, a similar process for manufacturing membrane vesicles by a mixing strategy is often employed in the laboratory. Lipids that are dissolved in nonpolar solvents spontaneously form vesicles when the solution is diluted with water; the exact conditions employed determine the size and other features of the vesicles. Such vesicles—sacs bounded by artificial membranes—have been extensively studied by membrane biologists; investigations of this kind have greatly extended our understanding of natural (i.e., real) membranes.

"Prebiotic" Chemical Reactions

If life originated in, or near, thermal springs, then thermophilic *Archaea*, long-term residents of such environments, may constitute a link with life's earliest evolution. It is possible that raw materials for early life may resemble those that underlie the ecology of contemporary thermophiles. It may therefore be instructive to think about the identity and origin of such materials. Clearly, the advent of life required the availability of such important small molecular precursors as amino acids, nucleotide bases, coenzymes, and sugars. Many of these served as precursors of polymers such as polypeptides (from amino acids) and RNA (from nucleotide bases, phosphate and sugars) that were formed by linking individual molecules by condensation reactions. It is clear that the manufacture of such precursors and their polymers was not identical to the emergence of life, but it was an absolute prerequisite.

Reactions leading to the synthesis of such precursor molecules before life was established are termed prebiotic reactions and a great deal of effort has been expended in studying them in the laboratory under conditions that attempt to imitate those on the early Earth. Moreover, the abiological synthesis of such compounds on the contemporary Earth cannot be excluded: modern metabolism may still be an extension of "prebiotic" chemical transformations. And, if these transformations are features, say, of vent chemistry, then thermophilic *Archaea* may the organisms of choice to reveal the connections.

Sources of Precursors

It is not intuitively obvious how small building blocks, such as amino acids or nucleotide bases, became available. These were not likely components of an early Earth that had been formed through accretion of bits and pieces of the solar cloud. Indeed, for many years, the inability to identify sources of the important precursor molecules seemed an insurmountable barrier to the scientific study of life's origin. This situation changed dramatically in 1953.

At that time, the notion prevailed that the "creation" of life only entailed making the precursors. It was felt that, when precursors became available, they would spontaneously come together in just the right way to finish the process. Therefore, (the idea went) if nature could make an amino acid, proteins would somehow form automatically, and all else would follow. This might sound a bit silly now, but it really isn't: we saw earlier that the subunits of complex biological entities, like membranes and virus coats, do accrete in this way—the principle is called self-assembly. But membranes and virus coats are uncomplicated when compared to cells. Thus, producing a living cell in this manner would be a tall order, similar to a tornado descending on a junk yard and producing, by random "self-assembly," a jet liner. Not at all in the cards.

Energy for Prebiotic Reactions: Heat

It is certain that the precursors must have been available before anything else could happen. And it turns out that such availability is chiefly a matter of energy: if energy were not required for the synthesis of these compounds, they would all form spontaneously—they would just be there. And it happens that conditions on the early Earth were such that possible energy sources were present in abundance. For one thing, heat energy was vividly apparent: the young Earth was extremely hot, owing to the gravitational compression that created it in the first place. Other sources of heat included radioactive decay in the Earth's core, certain geochemical reactions, and also the effects of sporadic impacts of large celestial objects, such as comets and meteors, with collisions occurring at a diminishing frequency as time went on. We saw that, during the first half-billion years of the Earth's existence, the surface temperature probably exceeded 100 degrees Centigrade, water was mostly present as steam, and there was extensive volcanic activity. Heat energy, the vibrational energy of molecules, is, in principle, capable of providing energy to drive synthetic reactions, but it unfortunately (for prebiotic reactions) also has the effect of increasing the rates of *all* chemical reactions. In other words, heat might indeed drive the synthesis of amino acids and even accelerate synthesis of small peptides formed from them. However, at the same time, it would increase the rates of peptide hydrolysis to the individual amino acids and the subsequent breakdown of individual amino acids into fragments.

Thus, it is likely that the high temperature of the early planet figured in the history of life, more as a source of instability than as a source of energy for synthetic reactions. However, one cannot help thinking that an ideal way for heat to be "used" in prebiotic synthesis would be to carry out the syntheses at a high temperature and then rapidly cool the products, so that they wouldn't subsequently break down as rapidly. Hot vent water mixing with a cold ocean might do nicely.

Prebiotic Chemistry from Electrical Discharges

It is likely that the earth of four billion years ago was a stormy place: temperature gradients produced high winds in the atmosphere and a combination of vulcanism and associated chemical gradients created impressive electrical storms. Energy in the form of electrical discharges was particularly plentiful. Therefore, when it also became evident that the early atmosphere was relatively reducing and that it probably contained such gases as carbon monoxide and carbon dioxide, as well as water, methane, nitrogen, but very little oxygen, the stage was set for an experimental approach to prebiotic chemistry.

Thus, in 1953, Stanley L. Miller published what must be called a "revolutionary" account of the use of a spark gap to energize a laboratory simulation of prebiotic chemistry. The spark was continuously fired in a sealed flask containing what was then considered an apt model of the early atmosphere. The ingredients included methane, ammonia, hydrogen, and water. After some days, the flask was discovered to contain significant amounts of some twenty-five different amino acids, as well as a number of other compounds of biological interest. About 15 percent of the carbon from the initial methane was recovered as identified compounds: clearly the syntheses were as efficient as they were versatile.

This experiment was a turning point, showing that prebiotic chemistry and, by extension, the origin of life, was a subject that could be studied experimentally. Thus, it became at once a subject for real science, not just speculation. The experiment revealed that natural explanations could account for the availability of precursor molecules and that plausible energy sources could be identified.

During the years since 1953, our ideas about the likely composition of the early atmosphere have changed. For one thing, it has become more probable that much of the atmosphere originated, not as had been formerly believed, by condensation of the solar nebula, but through "outgassing" from the molten interior of the earth. This has led to the view that there was probably more carbon oxide and less methane about than had been supposed—that the atmosphere was less reducing than formerly thought. When electrical discharge experiments were conducted with updated model atmospheres, the same amino acids and other biologically interesting compounds were obtained, but with considerably lower yields. It has been frequently pointed out that the yields are of relatively little importance, in view of the very long times available to do whatever chemistry was necessary.

Analogous prebiotic synthesis experiments have been carried out with energy sources other than electrical discharge. Radiation with ultraviolet light, electrons, or gamma rays has also led to a similar collection of amino, and other, acids. Heat (about 1000 degrees) with the gases absorbed on quartz or silica gel surfaces can do the trick as well. There has been considerable interest in experiments where gas mixtures are passed over the surface of metal oxide and silicate catalysts at about 300 degrees. It seems that all these approaches have plausible relationships to early Earth conditions as we understand them, and that all produce a wide and interesting selection of biological precursors. The product distributions both differ and overlap when the various methods are compared, but, between them, they account for many important molecules. These include most amino acids found in organisms, purine and pyrimidine nucleotide bases, fatty acid components of phospholipids, and a variety of sugars and metabolic intermediates. Clearly, these reactions could set the stage for a great deal of subsequent biology.

Delivery of Precursors by Meteorites

A number of years ago, it became evident that certain types of meteorites contained organic compounds that were identical to those found in living material. This discovery was most unexpected and led, initially, to some quite unrestrained speculation about life in remote parts of the universe and, later, to the more sober realization that nonbiological organic synthesis could, in fact, occur in space. Organic molecules were subsequently detected in comets and interstellar dust, so that their occurrence turned out to be widespread. Evidently, the original nebula, from which the solar system condensed, contained at least the raw materials of such compounds. And energy to drive the syntheses was apparently available, perhaps in the form of cosmic radiation. Moreover, the universe contains a preponderance of reduced atoms—that is, those that are able to donate electrons. In the case of the spark discharge experiments, we observed that availability of reduced atoms enhanced the synthetic capability of the system.

There is reason to believe that the processes that led to the formation of the earth also resulted in depletion of organic material. For example, one view has it that the earth moon system originated from an impact with sufficient energy to, among other things, destroy most organic molecules that might be present. But we saw that the first billion or so years of the Earth were punctuated by a daunting rain of hard objects, including meteorites, cometary fragments, and the occasional larger asteroid. An important effect of these impacts was the delivery of tons of organic compounds. One estimate is that, during the Earth's first 0.1 billion years, about 10^{22} kilograms of cometary material arrived. And cometary nuclei have been described as "dirty ice" with much of the dirt being, in fact, carbon. Remarkably, it turns out that comets may have contributed enough water (from the ice) to account for much, or all, of the mass of the oceans and enough carbon (from the "dirt") to yield much, or all, or the world's present biomass.

Spectroscopic studies of comets also reveal the presence of small molecules, such as carbon monoxide, hydrogen cyanide, formaldehyde and hydrogen sulfide. These are the same molecules that are thought to be intermediates in some of the reactions leading to amino acids and nucleotide bases in electrical discharge experiments. Larger molecules have also been observed in comets, possibly including the raw materials of nucleic acids. In addition, a class of meteorites, called carbonaceous chondrites, which appear to have been relatively protected during their history, contain a large number of different amino acids, many of which occur in living organisms. These interesting meteorites have also been observed to contain a good assortment of other biologically relevant compounds. Clearly, such celestial objects constitute a major potential source for the raw materials from which life might have arisen.

Syntheses in Thermal Vents

Organic chemists often have to heat things up to get their synthetic reactions to work, a matter of surmounting an "energy barrier" that lowers the rate of the reaction. For a similar reason, the heated water in terrestrial or marine thermal vents might be an excellent place for prebiotic organic chemistry. After all, vent water is rich in dissolved materials that might serve as appropriate reactions, and is also often replete with reducing (electron-donating) compounds, like hydrogen sulfide; we saw earlier that reduced compounds promote syntheses. Furthermore, some of the minerals commonly present, such as iron pyrites, are well known to be effective catalysts of the kinds of reactions we are talking about. For such reasons, there has been lively interest in thermal vents as prebiotic chemical reactor systems, and a number of experimental studies have been carried out.

Of course, an obvious problem with heating things up to make organic reactions go is the likely instability of the products at high temperatures. Stanley Miller, the pioneer in electrical discharge prebiotic synthesis, is also a strong critic of the hydrothermal origin approach, arguing that the lifetime of any interesting products is simply too low for net synthesis to amount to much. He, and others, have used the same argument against the recent, and evidently extravagant, claims of the discovery of microorganisms living at temperatures in excess of 250 degrees. It is now clear that such claims are indeed mistaken, and that the lifetimes, at so high a temperature, of such molecules as DNA, proteins, and ATP are simply too short for organisms to survive.

However, when one considers vents as venues for prebiotic chemical reactions, the matter turns out to be a bit more promising. Hydrothermal vents are, after all, flow systems from which extremely hot fluids issue, to be mixed rapidly with the cool ocean waters. Such flow systems offer the possibility that compounds could be synthesized in the hot interior of the system after which they would be rapidly cooled by mixing with seawater, and, in that fashion, stabilized. With suitable flow rates and residence times for components in the various parts of the system, some quite subtle and inventive chemistry becomes possible. This sort of scenario is particularly attractive when one discovers that many vents on the ocean floor are lined with clays, sulfides, and pyrites, all of which have been found to catalyze some of the reactions required. It seems that synthetic chemistry in hydrothermal vent systems, probably widespread features of an earlier earth, could well have played a major role in providing the raw materials for life.

Do *Archaea* Bring Us Messages from the Early Earth?

We can now return to the question of whether the *Archaea* resemble, in any important way, early cells, or can provide information about conditions that prevailed in the era when life arose. Does archaeal biology, for example, yield insight into the matter of the location where life originated? Recall that the most ancient lineages of the *Archaea* and the *Eubacteria* belong to those varieties that live at very high temperatures, often in modern-day vents. Moreover, archaeal membranes, with their stable ether bonds, suggest a hyperthermal origin; possession of ether lipids is probably a primitive character as it occurs in all *Archaea*, even in archaeal varieties that now live at moderate temperatures. And, of course, many *Archaea* are strictly anaerobic. Most thermophiles and all methanogens are harmed by oxygen and one is encouraged to believe that such an anaerobic way of life is primitive—that is, was retained from the prephotosynthetic era when oxygen was scarce. Indeed, it was probably particularly scarce in sulfide-rich hydrothermal water. It may also be suggested that halophilic *Archaea* retain comparable vestiges of their early history, that their impressive salt tolerance originated under the influence of the mineral content of the early vent waters where, as some would suppose, it all began.

Perhaps here is an additional clue: no *Archaea* are truly photosynthetic, and many, somewhat inexplicably, obtain their energy and carbon from such complex organic molecules as sugars, amino acids, and even proteins. For example, the thermophilic anaerobe *Pyrococcus*, can metabolize starch and protein, using sulfur as an oxidant. One wonders if this taste for organic carbon compounds harks back to their earlier availability via vent chemistry. On the other hand, it may be that the taste for organic carbon lends strength to the alternate supposition that early life was heterotrophic, originating in Darwin's shallow pond. One always does well to remain open-minded about such matters. Finally, the absence of chlorophyll and photosynthesis in the *Archaea* suggests that light energy did not, as many have proposed, serve as the basis for early life. Photosynthesis is restricted to invariably nonthermophilic *Eubacteria* and their descendants, the eukaryotic chloroplasts, and so could hardly have been a primitive feature of life. Thus, from a reading of the archaeal text, one is perhaps entitled to envision the first life as arising in heated water, without the benefit of either light or oxygen. Perhaps it will turn out, after all, that the *Archaea* are entirely worthy of their name.

Making a Living (Obtaining Energy)

Just as one sometimes identifies people by their means of livelihood, one can group microorganisms according to the way in which they manage their energy supply. Thus, photosynthetic bacteria obtain energy from sunlight, whereas most aerobic microbes obtain it from respiration. But, just as classification of people according to how they make a living may ignore their most important human attributes, so does grouping organisms according to energy source omit much of significance. For one thing, it says almost nothing about evolutionary relationships: usually, evolutionary classification (e.g., *via* RNA base sequences) and grouping according to energetics have very little connection. On the other hand, biological energetics does say a great deal about how organisms interact— how they fit into a larger ecological web of mutual influence. Also, the way that an organism handles energy is strongly conditioned by the character of its nonbiological environment, so that alteration of the environment can be closely linked to adaptive changes in energetic processes. For this reason, the study of biological energetics provides clues about how the biosphere, as a whole, has evolved.

Among the *Archaea*, there are creatures that obtain energy in quite unusual, even seemingly paradoxical, ways. Some halophiles obtain energy from light, but are not in the usual sense photosynthetic. Most methanogens obtain energy by reducing carbon dioxide to the natural gas methane, with hydrogen often serving as the reducing agent (source of electrons). Finally, thermophilic *Archaea* also approach matters of energy with style and originality, in ways that may provide insight into their long history. Thus, in the world of thermophiles, sulfur plays a paramount role in which it, and not oxygen, is the focus of respiration. This preference may reflect conditions on the young Earth, where sulfur was common and oxygen was not. In this fashion, the evolution of organisms and that of the Earth itself appear closely coupled.

What Is Energy?

In regard to energy, customary language can be misleading. One speaks of an archaeal cell "obtaining" or "transferring" energy, just as if energy were a substance, whereas we have known for over a century that it isn't. Before then, energy was considered "stuff"—perhaps another element, like carbon or oxygen. The demolition of this erroneous view was one of the significant achievements of nineteenth-century physics, but the idea of energy as "stuff" clearly persists in our language, making it a great deal harder to understand what energy really is. Therefore, to make sense of archaeal energetics, we should be clear about the meaning of energy in general. To this end, we turn first to several definitions.

Energy has a short and a long definition. And, remarkably, the short definition is actually correct. (This is fortunate because the long definition is very long—at least book length). Here is the short version: energy isn't a something and it isn't an event, but energy is that property of an event that allows it to occur. If that isn't perfectly clear, consider two key observations.

First, some things happen spontaneously in nature; others don't. A ball rolls spontaneously down a slope, not up. A chemical reaction proceeds spontaneously in one direction, but not the other. Heat will spontaneously flow from the hot end of a metal bar toward the cold end and not vice versa. And if one shakes an assembled jigsaw puzzle in a barrel, it becomes disorderly, whereas if one shakes the disordered pieces, even for a very long time, they don't reassemble themselves into the orderly completed puzzle. So some things happen spontaneously, others don't, and different explanations seem to underlie different cases. Gravity determines the direction of movement of a ball on a slope, while some statistical constraint prevents the puzzle from reassembling itself without laborious help. If an event is spontaneous, its reverse isn't (and vice versa). If the chemical A is spontaneously transformed into B, then the reverse reaction, B to A, doesn't "go." Thus, if the hydrolysis of an archaeal protein is spontaneous, the reverse reaction—its synthesis—cannot be.

There are also instances in which an event that isn't normally spontaneous can be made to occur by coupling it to a spontaneous one. For example, a locomotive can coast down a hill, but, in fact, it can also, with the use of steam, climb up it as well. In this case, the upward climb of the locomotive is coupled to the spontaneous expansion of water into steam, with the coupling being accomplished by an ingenious arrangement of pistons, cylinders, and other hardware. Now, our habit of language encourages us to say that "heat energy" is being transformed into "mechanical" energy—just as if energy was a thing. But it really isn't. We have detailed understanding of every step in the process, from heating the water, conducting the steam to the pistons, and making the great wheels turn, and any substance called "energy" is noticeably missing from the chain of causation.

Energy as Metaphor

Here is a productive way to think about energy. The word "energy" is a sort of shorthand, a metaphor even, that helps us organize our thinking about the rather complicated matter of spontaneity. There is a convention in this shorthand for describing things happening (or not happening). The convention has rules that go like this:

1. Events always occur in such a way that energy declines. (Thus, the change in energy for any event is negative.)
2. Events are additive and can form chains. And rule 1 applies to the sum of events in the chain.

3. Energy is also additive and different kinds of energy—heat energy, mechanical energy, electrical energy, even statistical energy—are interconvertible, and must add up.

Actually, when we say "energy" in rule 1, we really mean what specialists call free energy, defined as the form of energy that affects the spontaneity of events. Another way to say the same thing is that free energy is available to do work. Thus, "doing work" really means rendering an event spontaneous.

All cellular chemistry is governed by rule 2. The coupling of reactions can only result from the product of one reaction being a reactant in another. For reactions to affect one another in terms of energy or spontaneity, they must be linked in chains (or pathways). Then a spontaneous reaction can "push" or "pull" a nonspontaneous one; the reactions are coupled energetically. And evolution has devised effective strategies for linking reactions in life processes, coupling the spontaneous ones to the others. Often this amounts to inserting spontaneous steps into a chain of unfavorable reactions so that the entire chain become spontaneous.

Energy in the Biological World

Organisms require energy to live and, when they don't "obtain" it, they stop being alive. Putting it more accurately, living organisms must perform tasks that are not, by themselves, spontaneous. These tasks, which include the synthesis of macromolecules, must be linked to spontaneous processes. For instance, in *Sulfolobus*, spontaneous reduction of sulfur can be coupled, via a chain of reactions, to an energy-requiring process such as protein synthesis.

The central principle, then, is that biological processes can only be energetically coupled via such chains of events; to assert otherwise would constitute advocacy of magic. This principle explains why organisms cannot exploit all conceivable sources of energy: there must be the possibility of forming the required chains. For instance, thermophiles, like *Sulfolobus* live in the presence of abundant heat energy, but have no way of exploiting it, and resort to the spontaneity of certain reactions for their energy "supply." Many *Archaea* are exposed to sunlight, with its possibility for spontaneous photochemistry, such as the chemistry underlying photosynthesis. But only a few halophiles have the capability to link light absorption to cellular processes and, as we noted, no *Archaea* are truly photosynthetic.

Protein synthesis is a condensation that links amino acids; consequently, it is the reverse of protein hydrolysis. We saw that the hydrolysis is highly spontaneous; therefore synthesis isn't, and must somehow be linked to some spontaneous event. Evolution has evolved a cunning solution to this linkage problem. The condensation does not occur as a single reaction, but is partitioned into a chain of subsidiary reactions. And one of these is highly spontaneous, enough so to drive the whole chain. That spontaneous reaction is the transfer of a phosphate ion from adenosine triphosphate (abbreviated ATP) to some other molecule, an intermediate compound in the overall chain of events.

The Role of Adenosine Triphosphate

Evolution does not often waste a good solution that can be applied in multiple instances. Adenosine triphosphate (ATP) is utilized in the same manner to render a variety of cellular processes spontaneous. These include many synthetic reactions; in each case, ATP transfers a phosphate as a part of a chain of events. ATP also provides a spontaneous step in many other processes. For instance, it is used to move ions across a membrane against a concentration gradient and to drive muscular contraction. The pivotal importance of phosphate transfer from ATP has led to an economic analogy: it is a currency to be earned and expended. In fact, ATP does cycle in the fashion of money, produced as the result of various spontaneous processes

$$ADP + \text{PHOSPHATE} \rightarrow ATP + \text{WATER}$$

and then "spent" making nonspontaneous things occur

$$ATP + \text{WATER} \rightarrow ADP + \text{PHOSPHATE}$$

These two reactions form a closed loop and, in a well-ordered cell, the two are in perfect balance.

One should add that other nucleoside triphosphates, such as cytosine triphosphate (CTP), guanosine triphosphate (GTP), and even inorganic pyrophosphate (PPi), can serve as phosphate donors in the same fashion.

Spontaneous processes responsible for the synthesis of ATP and similar compounds are of three sorts: photosynthesis, cell respiration, and direct chemical phosphorylations. Photosynthesis is ATP synthesis coupled to electron transfer initiated by light absorption. Cell respiration is ATP synthesis coupled to electron transfer between two different atoms, a donor and acceptor. (Often the electrons end up reducing an oxygen or a sulfur atom—the acceptor.) And "direct chemical phosphorylations" are often coupled to electron transfer between compounds within a metabolic pathway. Fermentations are an example.

Energy Coupling at Membranes

Prokaryotic cell respiration and photosynthesis are localized in plasma membranes, with photosynthesis occurring only in *Eubacteria*. Electron transfer associated with these processes creates gradients of hydrogen ion (H^+) or, less often, sodium (Na^+) across the membranes. It is as if electron transfer "pumps" the ion across the membrane, producing ionic gradients, which are then "used" to synthesize ATP. The manner in which an ionic gradient can confer spontaneity on ATP synthesis—in effect, reversing ATP hydrolysis—is still a matter of some conjecture. But it is clear that an ATP synthase enzyme complex resides in the plasma membrane and that, in the presence of a suitable ionic gradient, the enzyme catalyzes ATP synthesis. Without the ion gradient, the enzyme catalyzes ATP hydrolysis and can create a gradient: the enzyme is, at the same time, an ATP synthase and an ionic pump; both activities are reversible and tightly coupled to one another.

"Harvesting" of Light by Halophilic *Archaea*

The revelation that halophilic *Archaea* use light as an energy source occurred through a combination of accident and insight. Walther Stoeckenius, then at Rockefeller University in New York, was led to study the halophiles because of having read what turned out to be an erroneous observation about their membranes. In the process of finding out that the observation was, in fact, wrong, he discovered how interesting halophiles were, and went on to describe a unique mechanism by which they use light energy to affect the spontaneity of cellular activities. This was in 1964, long before the *Archaea* were even a gleam in Carl Woese's eye and, in fact, before our current fluid mosaic view of membranes was completely established.

The (erroneous) paper in question reported that halophiles lacked the cell wall that encloses most bacteria, and that, at low salt concentrations, the plasma membrane breaks down into uniform subunits. At that time, membranes were believed to be assembled from such subunits and Stoeckenius thought that the halophiles would be good organisms in which to look for them. However, it was soon clear that halophiles had a perfectly good cell wall and that the supposed membrane subunits were actually cell-wall fragments. But it also became apparent that halophile membranes were composed of fractions with three distinct densities, and Stoeckenius and his colleagues set about studying the implications of this unusual composition.

Their investigation turned out to be exceedingly productive, but, before pursuing it, we should say more about the biology of halophilic *Archaea*. First, they grow best at a sodium chloride concentration of about 4.3 M (251 grams per liter), whereas its concentration in ordinary seawater is only about 0.6 M (35 grams per liter). Very few organisms can endure 4.3 M sodium chloride, which is why salt is often used as a food preservative. This singular ability of halophiles to defend themselves against such salty conditions is closely connected with their energy-transfer ability. Recall that energy can be stored in the form of ion gradients, which can then be used to synthesize ATP. Halophiles necessarily live with enormous gradients of sodium and other ions, and these strongly affect their options for making a living.

Most halophiles obtain energy by oxidizing amino acids or other organic compounds. In other words, they carry out cell respiration, transferring electrons from the amino acid, via an electron-transfer chain, to oxygen. There, the electrons reduce oxygen to yield water. This electron transfer is associated with movement of hydrogen ions across the plasma membrane in an outward direction, producing a hydrogen ion gradient. The ATP synthase in the membrane uses the gradient to manufacture ATP (transferring an inorganic phosphate to ADP). Also, some halophiles are able to use sulfur in place of oxygen as an electron acceptor, an ability that we found to be rather common among other archaeal organisms.

Electron transfer from an amino acid to oxygen is a highly spontaneous reaction. This reaction is coupled to hydrogen ion movement, causing the spontaneous creation of a hydrogen ion gradient, which leads to highly spontaneous ATP synthesis. As we saw, subsequent phosphate transfer from ATP can render other processes spontaneous as well.

The Halophile Membrane

Walther Stoeckenius observed that halophile membranes consisted of fractions with three different densities. This was learned by centrifuging broken cells in a density gradient, using a centrifuge tube containing a more dense solution at the bottom, and changing continuously to a less dense one at its top. When broken membranes are placed in the tube and the tube is centrifuged, each component ends by "floating" at a density corresponding to its own. In this manner, the Stoeckenius group obtained membrane fractions that were distinguished by three different densities and also different colors. A low-density fraction was reddish, a more dense fraction purple, and the most dense was yellow.

The situation became simpler when it turned out that plasma membrane was composed only of the red and purple fractions; the heavy yellow fraction consisted of fragments of the walls of gas vacuoles, internal sacs of gas that regulate cell buoyancy. The red color turned out to be mostly due to the presence of a chemical called bacterioruberin, one of a class of compounds that protects membranes against the harmful effects of light. On the other hand, the identity of the purple pigment was, initially, a complete mystery. It turned out that the solution to this riddle was an exceedingly fruitful one.

The Purple Membrane

The purple membrane fraction had a number of unusual properties. For one thing, it was homogeneous, composed of particles of uniform size. This was unexpected as membrane preparations tend to be quite random in their size distribution. Also, unusual for membranes, the fraction was about three-quarters protein and one-quarter lipid, in contrast to about equal distribution of the two in most cellular membranes. Such a high protein concentration usually indicates that a membrane is biochemically active (i.e., with a high enzyme content). It turned out that the purple membrane was very active indeed.

Another unusual feature: when the purple membrane was studied by X-ray diffraction, it exhibited a structure similar to crystallized protein. This indicated a large measure of regularity in its molecular structure, again, a situation quite unlike that occurring in most cellular membranes. Finally, the regularity was also evident in electron micrographs of the purple fraction, showing hexagonal geometry. And, interestingly, when intact, unfractionated membranes were examined, the purple membrane was distributed as a patchwork. The expressive term "purple patches" was thus coined and has been with us ever since. At this stage of events, there were two compelling questions. First, what molecule is responsible for the purple color and, second, what is that molecule's function? There is clearly a close connection between the answers to these two questions.

By 1967, Dr. Stoeckenius had moved from New York to the University of California at San Francisco, and his research associates included Dieter Oesterhelt and Allen Blaurock. X-ray diffraction, chemical analysis, and electron microscopy were employed to characterize purple patches more fully. It soon became clear that the patch contained only one kind of protein, with a molecular weight of around 26,000 and that three of these protein molecules, together with associated lipids, composed the fundamental building block of the membrane. The molecular weight was "typical" for a protein, but it was definitely not typical for membranes to contain only a single kind of protein. For example, the inner mitochondrial membrane contains more than one hundred different types of protein.

The Color Purple

It is possible to identify the pigment that gives the purple patches their distinctive color by obtaining its absorption spectrum, a graph of absorbance versus wavelength. Such a spectrum is often unique to a particular compound and thus provides a good fingerprint.

The absorption spectrum of the purple membrane from *Halobacterium halobium* was obtained by Oesterhelt and Stoeckenius using purified membranes from cells that had been broken up by removal of salt. Cells were harvested by centrifugation and disrupted by removal of the sodium chloride by dialysis. *Halobacterium* membranes require a high salt concentration for stability and fall apart when it is lowered. Dialysis amounts to placing the cell suspension in a cellophane bag, which is immersed in distilled water. Small molecules and ions like sodium chloride migrate across the cellophane into the water and thus are diluted away. Membrane fragments, proteins and nucleic acids too large to cross the cellophane, are left behind.

The purple membrane was then purified by centrifugation in a density gradient in a tube containing a high concentration of sucrose in the bottom ranging to a low concentration at the top. There, the membrane fragments and other odd bits from the disrupted cells were separated according to their density. The purple membrane, purified in this fashion, exhibited a characteristic hexagonal pattern when viewed in the electron microscope. The absorption spectrum of the purple membrane revealed two absorption peaks, one at about 280 and one at about 570 nanometers.* The 280 nanometer peak is in the ultraviolet region—beyond the range of the human eye—whereas the 570 nanometer peak is in the yellow-green region. If light is absorbed in that region, transmitted light appears purple.

The Stoeckenius team then treated the purple membrane with various chemicals, attempting, among other things, to extract the pigment into organic solvents or solutions of detergents. Such procedures generally led to bleaching of the 570

*Wavelength is expressed in units of distance; a nanometer abbreviated nm is 10^{-9} meters, or one-millionth of a millimeter.

nanometer peak, so that the purple color disappeared. It should be mentioned that the 280 nanometer peak is in the region where all proteins absorb light, and we know that the membrane is largely protein. Evidence began to accumulate suggesting that the visible color—the 570 nanometer peak— was due to a chromophore, a light-absorbing molecule bound to the single protein that was present. And, evidently, when the chromophore was removed from the protein, it lost the purple color. But what was that chromophore?

The Discovery of Bacteriorhodopsin

Given time, no doubt, someone would have stumbled over the identity of the mysterious purple chromophore, the source of the purple membrane's distinctive color. However, by a bit of good fortune, a Stoeckenius colleague, Dr. Blaurock, had formerly worked in a laboratory where the chemistry of animal vision was under investigation. Animal eyes detect light using a pigment consisting of a small molecule, retinal, bound to a protein, called an opsin; the combination is termed rhodopsin. Retinal, a derivative of vitamin A, absorbs light in the ultraviolet, at 280 mn. Thus, the Stoeckenius Blaurock Oesterhelt research team was aware of rhodopsin and its attributes, although there is no reason why anyone would expect an animal visual pigment to crop up in a bacterium, which is what *Halobacterium halobium* was still considered in 1967.

In fact, another observation had prepared their minds for that possibility: Stoeckenius had earlier noticed that *Halobacterium* cells were motile (by means of flagella) and could respond to changes in red light intensity by reversing their swimming direction. Thus, they could sense light, an ability shared by the animal retina, although it was rather far-fetched to expect the same sort of molecule to be involved. But, early in the history of vision research, the retinal pigment was named visual purple and purple pigments are not terribly common in the living world.

So Oesterhelt et al., considering the possibility that the purple membrane might contain something like rhodopsin, tried a variety of tests commonly used to identify the visual pigment. And, indeed, they found that the membrane color was due to a retinal molecule bound to a protein that very much resembled the animal opsin. The *Halobacterium* pigment was therefore given the name bacteriorhodopsin.

A key property was shared by bacteriorhodopsin and animal rhodopsin. In both instances, retinal absorbs light at about 370 nanometers, below the limit for human vision, so that pure retinal appears colorless. However, when it is bound to opsin, part of the retinal is internalized in the protein, shielded from the surrounding water. In this shielded state, the absorption peak is shifted far toward longer wavelengths—to about 570 nanometers—giving the complex its purple color. Moreover, when the complex is treated with a detergent, the shape of the complex is altered, the shielding abolished, and the absorption peak reverts to the near ultraviolet. So the purple color is bleached away.

Speaking of bleaching, visual rhodopsin exhibits the additional property of being bleached by light of the correct wavelength. Thus, when visual pigments are illuminated, they lose their color owing to a large change in their spectrum. Alternation of bleaching and restoration of color constitutes the primary event in visual excitation—the "visual cycle." Initially, such events were not observed with the *Halobacterium* system, making it difficult to see how light could be doing anything very useful for the cell. But, before long, conditions were found that allowed such bleaching to be observed: purple membrane suspended in a high-salt solution became colorless on addition of ether *and* illumination with visible light. This bleaching was reversible: color returned in the dark.

Later, it was found that the bleaching cycle could be observed without ether if the measurements were carried out at an extremely low temperature. It developed also that the transition from purple to colorless occurred by way of several intermediate states, identified by their characteristic, but transient, absorption peaks. For example, under just the right conditions, a peak at 412 nanometers would emerge, and then decay, as the cycle progressed.

Perhaps the most important observation of all was that light-induced bleaching is accompanied by release of hydrogen ions. Moreover, the subsequent recovery of color was accompanied by binding of hydrogen ions. Because pH is an inverse measure of hydrogen ion concentration, the binding and loss of hydrogen ion can be followed with an ordinary pH meter. In addition, the location of bacteriorhodopsin in the plasma membrane, together with the well-established role of hydrogen ion gradients in cellular energy, was highly suggestive: perhaps the bleaching cycle was part of a novel system for trapping energy.

Bacteriorhodopsin is a Hydrogen Ion Pump

Halobacterium halobium, and related halophiles that can make bacteriorhodopsin, do so actively only when light intensity is high and oxygen relatively low. These conditions are commonly realized in nature, as the organisms grow in very sunny locations and reach high densities, where they are likely to deplete the available oxygen. In other words, halophiles manufacture rhodopsin when there is plenty of light energy available and also when their ability to obtain energy through respiration is compromised.

Moreover, rhodopsin is embedded in the membrane with a uniform orientation, or "sidedness," such that, when hydrogen ions are released on illumination and bleaching, they are expelled only at the outer side of the membrane. Conversely, when rhodopsin regains its color, hydrogen ions approach it only from the inside surface of the membrane. Thus, a cycle of photochemical bleaching and recovery also results in a net outward movement of hydrogen ions. Therefore, the rhodopsin system is a hydrogen ion pump. Its pumping action can produce a hydrogen ion gradient and, as we have seen, such gradients can drive ATP synthesis.

Hydrogen ion gradients are pivotal in energy-related processes. In the case of *Halobacterium halobium*, a gradient can be formed either by the bacteriorhodopsin pump or by means of ordinary cell respiration. When oxygen is available, cells can respire, a process that creates a hydrogen ion gradient. Both cell respiration and the light-driven bacteriorhodopsin pump transfer electrons outward across the membrane. Therefore, their contributions to the gradient are additive. Moreover, the relative contributions of the two systems to the gradient are regulated by availability of light and oxygen. When there is plenty of oxygen, respiration prevails; when there isn't, the light-driven pump comes into its own, as long as there is adequate light intensity. Significantly, the amount of bacteriorhodopsin in the plasma membrane is diminished by oxygen and increased by adequate light intensity.

The hydrogen ion gradient is "used" to promote the spontaneity of cellular processes in several ways. First, the membrane contains an ATP synthase that couples ATP production to the inward movement of hydrogen ions. The gradient can thus be used to manufacture ATP. Because all the reactions involved in making ATP are reversible, the breakdown of ATP is associated with an outward movement of hydrogen ions. Thus, ATP can be used to produce the hydrogen ion gradient as well: it is a two-way street.

In addition, the hydrogen ion gradient can provide energy for the transport of amino acids and other useful compounds. If energy is "provided," these can be transported into cells against their concentration gradient—from a low concentration toward a higher one. In most *Archaea* (and also *Eubacteria*), transport of amino acids, sugars, and other nutrients is directly coupled in this fashion to hydrogen ion gradients. However, in certain halophiles, such as *Halobacterium*, transport of such molecules is linked directly to sodium gradients and only indirectly to those of hydrogen ions. This is possible because of the interconvertability of hydrogen and sodium ion gradients: a sodium gradient can supply energy for the formation of a hydrogen ion gradient and *vice versa*.

Conversion of energy from one form (e.g., light) to another (e.g., an ion gradient) in engineering language is termed "energy transduction." Bacteriorhodopsin is therefore an energy transducer par excellence, and it long ago attracted the attention of those who sought to apply it to biotechnology or electronics. Bacteriorhodopsin appeared to have some special advantages as a transducer, including its great stability (for a protein), functioning at very high salt concentrations and high temperatures, well over 100 degrees. Its photochemistry is also very rapid and durable: unlike many dyes, it doesn't use absorbed light energy to self-destruct. As an example, Russian scientists have been in the forefront of exploring the use of the pigment in photoelectric cells and imaging devices, but practical success has thus far eluded them.

A Second Light-driven Ion Pump: Halorhodopsin

Halobacterium and its relatives live in concentrated salt solutions. Being surrounded by large quantities of sodium and chloride ions, they must continuously expel these ions to keep their interior cytoplasm at levels suitable for life. The sodium part is easy. There is a transport protein in the membrane that exchanges sodium for hydrogen ions, so that the hydrogen ion gradient, with the high concentration on the outside, drives sodium expulsion. Notice that this exchange can convert a hydrogen ion gradient into a sodium gradient, which is then used, as we saw, to promote the uptake of things like amino acids and sugars against their concentration gradients. If one thinks about it, it is quite reasonable for a halophile, usually surrounded by high sodium, to employ a sodium gradient for uptake of other molecules. All that is required is a way of keeping the internal sodium concentration down, and sodium-hydrogen ion exchange does the trick nicely.

The chloride part is also easy—too easy, in fact. To see why this is so, we must introduce the notion of a membrane potential. First, a hydrogen ion carries a positive charge. Therefore, its flow, whether due to either respiration or bacteriorhodopsin pumping, is associated with the outward movement of positive charge: the inside of the membrane becomes relatively negative and the outside relatively positive. Such charge displacement across a membrane, with one side being more positive (or negative) than the other, is called a membrane potential. A voltmeter would register a voltage across the membrane, amounting to perhaps a tenth of a volt. Such membrane potentials are extremely important in biology, among other things, forming the basis of nerve transmission.

Membrane potentials also work in conjunction with ion gradients, enhancing their ability to affect spontaneity of processes like amino acid transport and ATP synthesis. Here is how they do it. It is an often-stated rule in physics that "like charges repel one another; unlike charges attract." Thus, two positive ions experience a force that drives them apart, whereas a positive ion (cation) is attracted to a negative ion (anion). And if a membrane exhibits a potential difference, an ion will be driven toward the oppositely charged side. For this reason, negatively charged chloride is driven out of halophile cells, away from the negative interior, and toward the positive outside of the membrane. Therefore, the cell doesn't have a problem getting rid of internal chloride, but, in fact, does it too well. Unfortunately, the cell requires chloride in its cytoplasm to balance the positive potassium ions, which halophiles require at fairly high concentrations for normal enzyme activity. So these organisms have actually evolved a second photochemical pump, which takes up chloride from the surrounding environment and so keeps things in balance.

The chloride pump is known as halorhodopsin and is present in the *Halobacterium* membrane in much smaller amounts than bacteriorhodpsin. The function of halorhodopsin was discovered by Janos Lanyi, who works in Irvine, California. Halorhodopsin is similar to bacteriorhodopsin in containing retinal as its light-harvesting chromophore, but differs in a number of details, so that it can be easily differentiated from the latter. Notice that bacteriorhodopsin transports the positive sodium in an outward direction, whereas halorhodopsin transports the negative chloride in an inward direction. Therefore, both contribute to a membrane potential with the same polarity—with the inside of the membrane negative with respect to the outside. In this sense, both affect chemical spontaneity, otherwise known as energy transfer, in the same manner.

So we discover that halophilic *Archaea*, including *Halobacterium halobium*, can obtain energy by the ordinary process of cell respiration, but also can use two light-driven pumps, one that transports hydrogen ion and one that transports chloride. Light harvesting by these pumps is based on rhodopsin, a protein-retinal complex otherwise found only in the animal retina. It is likely that the occurrence of such a similar protein in such remote representatives of the living world is an example of convergent evolution, independent evolution of the same solution to a common problem from quite unconnected precursors.

Energy Management in Thermophiles

One might imagine that thermophilic *Archaea*, living so intimately with more thermal energy than most organisms can tolerate, would exploit their situation and somehow use the heat for their own bioenergetic purposes. However, in most respects they do not: their strategy for energy coupling appears little affected by high ambient temperatures, at least in any positive way. Thus, *Sulfolobus* grows in the midst of sufficient geothermal energy to gladden the heart of any Icelander, but apart from the adaptations that render their proteins resistant to heat denaturation, is relatively unaffected. Of course, it is a temperature gradient, and not heat energy per se, that permits energy transfer—the reason why thermal electrical generating plants require cooling towers or a cool river nearby. Cells are simply too small to allow for significant temperature gradients within their scale of dimensions. Moreover, temperatures of around 100 degrees in thermophilic environments are really not sufficiently higher than those of an ordinary environment to alter reaction equilibria significantly and introduce spontaneity where none existed.

So, in fact, thermophiles make their living pretty much the way that most aerobic prokaryotic organisms do. They carry out cellular respiration, passing electrons from molecules like sugars and amino acids to oxygen or, often, sulfur. And as a result of respiration, they pump hydrogen ions out of their cytoplasm, producing a gradient that can be used for ATP synthesis or transport of other ions and molecules. In this, they resemble most representatives of all three domains.

Breathing Sulfur

Thermophilic *Archaea* do differ from other microorganisms in aspects of their cellular respiration. In all forms of respiration, electrons are transferred from an initial donor to a terminal acceptor (such as oxygen) *via* the respiratory chain of electron carriers; such electron transfer is coupled to the formation of an ion gradient. In this process, sulfur can also serve as a terminal electron acceptor—and sometimes an original donor too—and such sulfur reductions and oxidations are almost universal features of thermophilic *Archaea*. Indeed, most thermophilic *Archaea* are obligate anaerobes, growing only in the absence of oxygen. Therefore, they often employ sulfur instead as their respiratory electron acceptor. A few, like *Sulfolobus* and the similar *Acidianus*, are facultative anaerobes: they can grow either in the presence or absence of oxygen. Both can use either oxygen or sulfur as respiratory electron acceptors, as availability permits. It should be added that *Archaea* do not have a monopoly on sulfur oxidation-reduction reactions: many *Eubacteria* engage in this sort of activity as well. But the practice is nearly universal among the thermophilic *Archaea*—really, a hallmark of the group.

Sulfur occurs in nature, either as the element or in the form of compounds. Often it is elemental sulfur that serves as the electron acceptor. When electrons reduce elemental sulfur, the sulfur atom acquires two hydrogen ions, producing $H_2.S$; this is hydrogen sulfide and smells like rotten eggs. Notice the formal similarity of respiratory sulfur reduction (which yields H_2S) and oxygen reduction (which yields H_2O, water). In both cases, an atom is reduced with two electrons and two hydrogens are required for completion of the product. As sulfur and oxygen play such similar roles in respiration, one often humorously speaks of the archaeal habit of "breathing sulfur." In some instances, not elemental sulfur, but compounds or ions containing sulfur serve as electron acceptors. For example, the interesting thermophile, *Archaeoglobus*, a resident of marine hydrothermal vents, employs the anion, sulfate (SO_4^{--}) as an electron acceptor, reducing it to hydrogen sulfide plus carbon dioxide. This is unusual, but *Archaeoglobus* is, in other respects, an unusual beast, combining many of the properties of a "mainstream" thermophilic archaeon with those of a methanogen.

Many thermophilic *Archaea*, whether using sulfur or oxygen as terminal electron acceptor, employ organic compounds—compounds containing both carbon and hydrogen—as respiratory electron donors. These include relatively simple sugars, amino acids, ethanol, and various metabolic intermediates, small molecules that are normal constituents of cells. They also include more complex items: peptides composed of several amino acids, as well as polysaccharides, which are polymers of simple sugars. In many cases, such organic compounds represent absolute requirements for the isolation and growth of these organisms. Recall that yeast extract—a mixture of just about every organic molecule under the sun—is used for the isolation of *Archaea* such as *Sulfolobus acidocaldarius*. Such requirements for organic nutrients are rather mysterious in organisms that live in austere hydrothermal waters, where the presence of such compounds would be unexpected. One might well ask how it is possible that these creatures have evolved the enzymes required to utilize such molecules.

In addition, many thermophilic *Archaea* employ inorganic electron donors in their respiration. For example, *Archaeoglobus* uses hydrogen gas as an electron donor and electrons from it are passed by way of the respiratory chain to its acceptor, sulfate. Similarly, *Pyrodictium*, the marine hydrothermal organism with an optimum growth temperature exceeding 105 degrees, can pass electrons from hydrogen gas to elemental sulfur, producing hydrogen sulfide.

$$H_2 + S_{(elemental)} \leftrightarrow H_2S$$

In a similar fashion, *Sulfolobus*, growing in the presence of hydrogen and oxygen, can use the former to reduce the latter, again with the electrons being passed along a respiratory chain. The product of this reaction is water.

$$2H_2 + O_2 \leftrightarrow 2H_2O$$

Finally, let us return to the matter of the centrality of sulfur in thermophilic archaeal respiration. The organisms in question occur naturally only in hot places, such as geothermal springs, either on land or on the seabed. The concentrations of dissolved material in such water reflect the geology of the area, the exact path of the water through geological structures, and any biological activity due to resident organisms. In fact, many geothermally heated springs are rich in sulfur, either in the elemental form or as hydrogen sulfide. Often such waters are also rich in sulfuric acid, H_2SO_4, which is a product of the microbial oxidation of the elemental sulfur or hydrogen sulfide. Such biologically produced sulfuric acid accounts for many hot springs being so acidic, and, therefore, for the observations that many thermophilic *Archaea* are also acidophiles.

Fools' Gold

Elemental sulfur is a solid and in hot springs it often occurs as a suspension of small grains. If one examines such particles with a microscope, cells of thermophiles such as *Sulfolobus* can often be observed, clinging to their surface, presumably caught in the act of using the solid sulfur to respire.

Sulfur-rich hot springs often also contain considerable iron, some of which may be in the form of grains of insoluble iron pyrite. This mineral exhibits a gold-colored metallic surface and is appropriately known as fools' gold. Its chemical formula is FeS_2. When grains of it are obtained from hot springs and observed with a microscope, one notices cells of *Sulfolobus* clinging to the grains. Again, the cells have apparently been caught in the act of doing chemistry on the solid surface, this time reducing pyrite to ferrous sulfide, employing it, in place of sulfur, as an electron acceptor. This observation reminds one of the speculation that early life could have been confined to solid surfaces (such as that of pyrite). According to this view, the mineral surface would provide not only a place to adhere, but a chemical cafeteria replete with respiratory and other metabolic opportunities.

In the context of microorganisms inhabiting mineral surfaces, it may be added that bacteria have also been observed on the surfaces of microscopic gold "nuggets" from the sea bottom. (This is real—not fools'—gold.) It seems likely that the bacteria are responsible for the chemical reactions that lead to the precipitation of the gold in solid form.

Archaea Even Breathe Iron

Just as sulfur serves as a respiratory electron acceptor for many thermophilic *Archaea*, so does the oxidized (ferric) form of iron. If anything, the use of iron in respiration is even more widespread among the *Archaea* than that of sulfur, and there is a growing suspicion that iron-based respiration was a prominent feature of very early life. Both iron and sulfur in their reduced (electron-rich) forms commonly occur in hydrothermal water, so that both were probably widely available in the early Earth and there is ample geological evidence that biological iron reduction was occurring at a very early date. For that reason, it is significant that most thermophilic microbes, both archaeal and eubacterial, that reduce sulfur in the course of their respiration also reduce ferric iron. Indeed, some *Archaea* reduce iron, but not sulfur, and some, previously believed to be incapable of any type of respiration, turn out to be iron reducers.

The likely prominence of iron respiration in early microbial evolution is emphasized by its almost universal occurrence among those organisms that are considered least evolved from the last common ancestor, organisms belonging to such archaeal genera as *Pyrodictium, Archaeoglobus,* and *Methanococcus,* as well as the eubacterial *Thermotoga.* Finally, just as microbes can grow on, and utilize, the surface of solid minerals such as iron pyrite, so can many *Archaea* reduce iron in the solid form of iron oxide (rust), converting it to the mineral, magnetite. Widespread magnetite deposits of great geological age probably reflect such microbial respiration and support the idea of iron as a significant electron acceptor in early forms of life.

And here is food for thought: practically all organisms employ intracellular iron and sulfur in their respiratory electron transfer pathways. Iron is oxidized and reduced as the active part of the cytochromes that are electron transfer proteins, and sulfur plays a similar role in other oxidation-reduction proteins and certain coenzymes. It is as if, during the course of evolution, electron transfer by these elements had been incorporated from the environment into the cell in association with specific, newly evolved proteins. In this fashion, the genesis of much of cellular chemistry may be viewed as the internalization of processes from the environment into the evolving cell.

Leaky Membranes

It is commonly observed that membranes become leaky to hydrogen ions at elevated temperatures. In fact, thermophilic *Archaea* and *Eubacteria,* at their customary, high environmental temperatures, are much more permeable to hydrogen ion than their more normal relatives that live at lower temperatures. A leak of this kind represents a short circuit—an unproductive collapse of the gradient that is not coupled to ATP synthesis or nutrient transport. For this reason, energy metabolism in the thermophiles is relatively inefficient and they must necessarily respire at a higher-than-ordinary rate to make enough ATP for a normal livelihood.

ATP Synthases of Thermophiles

When cells use the energy stored in the form of ion gradients to make ATP, they employ an enzyme complex called an ATP synthase. Because these systems play such a pivotal role in energy transfer, their evolution is of great interest and they have been carefully examined in the *Archaea*. Unfortunately, the situation turns out to be a bit complicated. For one thing, the thermophile *Sulfolobus acidocaldarius* contains not one, but several, distinct ATP synthases. These can be told apart by their different sensitivities to specific inhibitors, their different susceptibilities to activation by ions such as sulfate, and their different pH optima. For instance, one of them operates best at a fairly neutral pH (6.5) whereas another exhibits a quite acidic optimum (2.5). In at least one case, the synthase only catalyzes the formation of ATP as a sideline: its real function appears to be the manufacture of inorganic pyrophosphate.

And the situation is even more complicated than that: each synthase is composed of several different protein subunits, so that two different ones can have some, but not all, subunits in common. And while most of them couple ATP (or pyrophosphate) synthesis to hydrogen ion movement, at least one synthase from *Thermoplasma acidophilum*, the denizen of hot coal mine wastes, appears to utilize fluxes of the anion, sulfate, instead.

In making sense of this, it is helpful to know that in the living world, there are three distinct lines of synthase evolution leading to three classes of synthase complexes. These differ in their composition, chemical mechanism, and inhibitor sensitivity and are exemplified by

1. The sodium-potassium ATP synthase of plasma membranes of eukaryotic cells. These are also called P-ATP synthases to indicate that the enzyme is itself phosphorylated during the catalytic process.
2. The hydrogen ion synthase found in the plasma membranes of most prokaryotic cells, and also mitochondria and chloroplasts. This is called the F_1 class of synthases, named for one of its subunits, the one in which the catalysis actually occurs.
3. The vacuolar synthases, found in membranes that surround eukaryotic vacuoles, as well as in plasma mem-

branes of various prokaryotes. This is similar to the F_1 class, but with differences in subunit structure and inhibitor sensitivity.

It appears that the ATP synthases of thermophilic *Archaea* fall in the last two classes and share none of the features of the first. In fact, some students of the topic consider the *Archaea* to have their own special "A-type" synthase and use the word chimera to describe it (a chimera is derived from bits and pieces of both antecedents).

Such protein shuffling is consistent with other things that we know. For example, the synthases consist of catalytic and noncatalytic subunits; the site where ATP is made is on the catalytic one. The noncatalytic subunits may have been derived from the catalytic ones during the course of early evolution by gene duplication, a process that can be detected in the form of large repeating amino acid sequences in proteins. In fact, the vacuolar synthase probably arose from F_1 proteins by the same sort of process. This origin is evident in the size of their subunits, which have twice the molecular weights of the corresponding ones in the F_1 synthases. So it seems that the archaeal, F_1, and vacuolar-type systems are members of an intricately connected family, possibly descendants of a single common ancestor, but also products of a good deal of sequence trading and duplication.

Finally, at least one eubacterium, *Thermus thermophilus*, has a chimeric archaeal synthase and at least one *Archaea, Methanosarcina barkeri*, has a F_1 synthase, the sort normally found in *Eubacteria*. In such instances, one must remember that prokaryotes possess a general ability to exchange genes and these anomalies may reflect such "lateral transfer." Note also that both *Thermus* and *Methanosarcina* are particularly ancient representatives of their respective domains, and so might be survivors from a time when synthase evolution was still relatively fluid.

The Wonders of Methanogenesis

Methanogenic *Archaea* cannot tolerate oxygen so that their methane (CH_4) production is a strictly anaerobic process. Methanogenesis leads to ATP synthesis, clearly the point of the whole exercise. Like other modes of ATP synthesis, methanogenesis proceeds by means of electron transfer—a sequence of oxidation-reduction reactions. As in the case of ordinary respiration, this electron transfer gives rise to a hydrogen ion gradient and the gradient is (predictably) used to make ATP. So, in a sense, methanogenesis is a form of anaerobic cellular respiration in which the final electron acceptor is a carbon compound that can be reduced to yield methane.

Respiratory electron transfer requires electron donors and acceptors. In the simplest case of methanogenesis, the electron donor is hydrogen gas, the acceptor is carbon dioxide, and the net reaction is

$$CO_2 + 4\ H_2 \rightarrow CH_4 + 2\ H_2O$$

In other instances, the same molecule can be *both* donor and acceptor, as in the case of methanol yielding methane plus carbon dioxide and water

$$4\ CH_3OH \rightarrow 3\ CH_4 + CO_2 + 2\ H_2O$$

Other precursors of methane include carbon monoxide, various compounds containing methyl ($CH_3 -$) groups, as well as formic and acetic acids. To keep things relatively simple, we will stick to one example, the hydrogen and carbon dioxide reaction shown previously.

First, a word about raw materials. As a rule, methanogenic *Archaea* occur in three kinds of anaerobic habitat: bodies of marine and fresh water, the digestive systems of ruminant animals, and, sometimes, hydrothermal springs, both fresh and marine. In the first two locations, carbon dioxide, carbon monoxide, hydrogen, methanol, and the rest are likely to be of biological origin, the products of other microorganisms. In hydrothermal effluents, geological processes play a major role; all methanogenic *Archaea* thus far isolated from marine hydrothermal vents use geochemically produced hydrogen. Contributions from microbes in adjacent sediments do not appear important.

So far, everything seems familiar: the reaction in question is reminiscent of *Sulfolobus* using hydrogen to reduce elemental sulfur to hydrogen sulfide. However, the mechanisms and cofactors employed are very different. Cofactors, also termed coenzymes, are small molecules that assist in specific metabolic tasks, such as electron transfer or the transfer of a group of atoms. In organisms other than methanogens, coenzymes tend to be universal, with virtually all organisms using a coenzyme abbreviated NAD to transfer electrons in cellular oxidations. Similarly, the coenzyme NADPH carries out reductions in synthetic processes, and another coenzyme called thiamine pyrophosphate assists in carboxyl group transfers.

Methanogenesis is shockingly idiosyncratic, however, and employs a bewildering collection of cofactors in its reactions. And methanogens often contain high concentrations of these compounds, whose presence can be used to identify them. For instance, coenzyme F_{420}, an electron carrier, is fluorescent and its blue-green emission is a good test of the presence of methanogenic *Archaea*. This coenzyme is also found in cells of the thermophile, *Archaeoglobus*, which, it will be recalled, was something of a missing link between thermophiles and methanogens and can, in fact, make modest quantities of methane.

Coenzyme F_{430} does not fluoresce the way F_{420} does, but it contains a nickel atom; the presence of this trace element is therefore also a useful way to identify methanogens. All methanogens contain nickel and require it for growth.

Here is how methanogenesis works in the simplest case when hydrogen and carbon dioxide are the electron donor and acceptor, respectively. The net process transforms a carbon with two oxygens into a carbon with no oxygens, but four hydrogens. This is a chemical reduction, being the addition of electrons plus hydrogens to an atom. In the first step of the process, carbon dioxide binds to a coenzyme, methanofuran, and is reduced with electrons from hydrogen. At this point, the carbon has lost an oxygen and gained one hydrogen. Next, the carbon is transferred to a derivative of methanopterin, a second coenzyme, where, in two steps, it loses its remaining oxygen and gains two more hydrogens. The carbon is now that of a methyl group (CH_3).

The Final Step

The final step is the transfer of this methyl group to coenzyme M—a third cofactor unique to methanogenesis—where the group is further reduced to methane (CH_4). When the methane is released, coenzyme M is free to react again, continuing the process. This final step, the only one directly responsible for energy transfer, entails the reduction of a methyl group to form methane. This reduction occurs by the transfer of electrons from yet another coenzyme, HS–HTP, to the methyl group *via* coenzyme F_{430}. Coenzyme F_{420} plays an analogous electron-carrying role in an earlier stage of CO_2 reduction. In all, six cofactors, unique to methanogenesis, carry electrons or carbon atoms in the process.

The net process of reducing carbon dioxide to form methane is extremely spontaneous, which is another way of saying that considerable useful energy is associated with it. That energy is employed to make ATP, which, by serving as a common intermediate, can contribute to the spontaneity of other processes. Methanogenesis can be the sole source of ATP for the *Archaea* that have that capability. And, as one has come to expect, hydrogen ions constitute the link between methanogenesis and ATP production: that final reduction of a methyl group to methane is associated with creation of a hydrogen ion gradient, with hydrogen ions being transferred outward, across the plasma membrane. Then, ATP is synthesized by a hydrogen ion-conducting ATP synthase of the sort we have encountered before.

One is entitled to ask how this final electron transfer reaction can lead to such coordinated movement of hydrogen ions, the same question that applies to the generation of ion fluxes in cellular respiration, photosynthesis, and rhodopsin pumping. In fact, the same principle underlies all these processes: the proteins responsible for electron transfer are oriented in the plasma membrane in such a way that reductions (electron and hydrogen donation) tend to be directed toward one side of the membrane and oxidations (electron and hydrogen acceptance) toward the other. In the simplest case, hydrogen ions are produced on one side and consumed on the other, so that there occurs an *apparent* flux, although a *particular* ion may not actually cross the membrane.

This notion of oriented oxidation-reduction reactions is at the heart of the chemiosmotic theory of energy coupling, first proposed by Peter Mitchell in about 1960, and only widely accepted about two decades later—about when the evolutionary place of the *Archaea* was becoming recognized. His theory envisions ion gradients as intermediates in energy transfer in both photosynthesis and cell respiration, explaining why both processes always occur at membranes. Detailed knowledge of the structures of methanogenic proteins supports such a directed path for hydrogens and electrons and provides strong support for the chemiosmotic principle.

8

Archaea as Ancestors

Here is a paradox: *Archaea* are undeniably prokaryotic, yet they exhibit many eukaryotic features concerning the manner in which they manage genetic information. Indeed, their genetic systems appear to include a rather complicated mix of prokaryotic and eukaryotic strategies for information handling. How is this situation consistent with what we know of their evolutionary history?

Archaea and *Eubacteria* branched apart early in the course of evolution. Their divergence, occurring well over two billion years ago, can be called the first major event in the history of life after its origin; only much later did *Eukarya* appear on the scene. But, on examination of their genetic systems—the ways that they organize and express the information encoded in their DNA—*Archaea* often much more closely resemble eukaryotic organisms, those greatly more complex organisms whose cells contain nuclei and other membrane-bounded organelles. At the same time, archaeal cells have some features in common with those of the prokaryotic *Eubacteria*. So which is it? Are the real archaeal affinities with the other prokaryotes or with the more advanced eukaryotes? The answer turns out to be "both," so we must now ask how such a thing is possible.

The molecular biologist W. Ford Doolittle and two coauthors offered a succinct, albeit preliminary, answer to this question when they entitled a 1994 paper "Archaebacterial genomes: eubacterial form and eukaryotic content." To be specific, *Archaea* and *Eubacteria* are alike in usually carrying most of their genetic information on a single, circular chromosome, a giant DNA molecule without free ends. But the two domains are distinctly different in some of the ways that they organize genes on that chromosome, as well as the way in which they express the genetic information through protein synthesis. In these areas, the *Archaea* do things more in a manner resembling the ways of eukaryotic organisms. Again, we must ask how this is possible. Or, to put it differently, can this seemingly ambiguous situation tell us anything helpful about the place of the *Archaea* in the evolutionary scheme of things?

Genes and Gene Expression

The genetic material of cells is composed of DNA that carries information specifying amino acid sequences of all proteins of the cell. It appears that the ability to construct proteins correctly is all the information that the cell requires: given the synthesis of the necessary proteins in the proper amounts, everything else follows, and the cell exhibits properties appropriate to its character and function. The prokaryotic genome, the entirety of an organism's genetic material, most commonly consists of a single chromosome that contains large numbers of genes (for instance, the single chromosome of *Escherichia coli* contains about four thousand of them), and each gene specifies the amino acid sequence of a single protein.

Protein synthesis proceeds via the formation of a messenger RNA (mRNA) intermediate; this, in turn, provides the direct pattern for the amino acid sequence. The information is transcribed from the DNA sequence language to that of the mRNA according to rules of complementarity (base pairing). This process of mRNA synthesis is catalyzed by an enzyme complex called RNA polymerase. The final stage of making the protein is called translation; it reads the information of the mRNA sequence to specify the order of the protein's amino acids. Here the mRNA sequence is "read" by transfer RNA molecules, which act as adaptor molecules, enabling the amino acids to find their proper place on the mRNA blueprint. The amino acids are then chemically linked to yield the protein product.

This final event is catalyzed by enzymes and requires energy in the form of ATP.

The Archaeal Chromosome

Most of an archaeal genome consists of a single, large, circular chromosome composed of double-stranded DNA. This is the usual prokaryotic arrangement, occurring in bacteria as well. For example, the circular chromosome of *Sulfolobus acidocaldarius* contains about 3,050,000 nucleotide bases, more commonly expressed as either 3,050 kilobases (Kb) or 3.05 megabases (Mb). This size is fairly typical of both *Archaea* and *Eubacteria*. The archaeal genomes that have been extensively sequenced thus far range from about 1660 Kilobases (*Methanococcus jannaschii*) to about 4000 Kilobases (*Halobacterium salinarium*).

It has been noted that the *Thermoplasma* genome, which has not yet been fully sequenced, is quite small—only about 1700 Kilobases. In this respect, as well as in features of its cell structure, including lack of a cell wall, *Thermoplasma* closely resembles the eubacterial mycoplasmas, some of which are based on genomes as small as 1300 Kilobases. However, on the basis of rRNA evidence, the two occupy remote places in their respective two domains: evolutionarily speaking, they are quite unrelated.

Archaeal genes are disposed on the chromosome in a manner similar to that in *Eubacteria* and rather unlike that in eukaryotic cells. For one thing, archaeal and eubacterial genes are often collected together in operons. These are clusters of several genes that are associated with a single function and expressed and regulated as a unit. Thus, genes that encode amino acid sequences of several enzymes in a particular biochemical pathway are often members of the same operon. When the product of the pathway is required by the cell, the genes are transcribed as a unit to produce the mRNA required to encode all the enzymes. Interestingly, archaeal operons often resemble in detail the corresponding eubacterial ones, with the genes being arranged in the same order on the chromosome.

But although the genes are often clustered in the same order, there is frequently a distinct difference in operon man-
agement. For one thing, there tends to be more autonomy
among genes in the archaeal clusters than in eubacterial ones,
with greater opportunity for individual regulation. Commonly,
portions of the archaeal operons are differentially expressed,
while in the eubacterial operons, the situation is more likely to
be all or none. On the other hand, when comparing archaeal
and eubacterial genomes, one often observes similarities in the
locations of operons on the chromosome. For example, many
operons are adjacent to the same neighboring operons when
one compares the eubacterium, *Escherichia coli*, with an archaeal
methanogen such as *Methanococcus vannielii*. In that respect,
similarities extend to the longer-range structure of the chromosomes in the two domains.

There is another significant commonality between *Archaea*
and *Eubacteria*: in both groups, usually no dead space occurs
within genes. In contrast, eukaryotic genes often intersperse
regions that are active in encoding amino acid sequences with
those that are not. The active regions—those that are expressed in protein synthesis—are called exons. The intervening
sequences that do not encode protein sequences are termed
introns. Cells get rid of the non-expressed (intron) information
by allowing the DNA to be transcribed to form the corresponding mRNA, after which the intron portions are excised by specific enzymes and the exon (informationally active) regions are
spliced together. In this fashion, a larger primary mRNA is
pruned to yield the smaller mRNA that actually constitutes the
pattern for ordering amino acids in protein synthesis.

However, there is an important exception to the no-intron
generalization: *Archaea* do have introns in those genes that
carry sequence information for making ribosomal and transfer
RNA. In this respect, they diverge from the *Eubacteria* and resemble the *Eukarya*, which have such introns too. On the other
hand, some extremely primitive eukaryotic organisms lack introns, so that the evolutionary picture is not completely clear-
cut. At present, the evolutionary origin of introns is the focus
of considerable debate, with little consensus about whether introns entered the eukaryotic picture before, or after, the origin
of the nucleus. The genesis of introns is unclear in another
sense: introns may either represent genetic ''junk'' that has

arisen in the course of evolution, perhaps parts of genes that have gone out of use. Or they may have originated as ribozymes, catalytic RNA playing an important role in RNA repair. Such repair activity has been observed in intron RNA and is believed to be extremely ancient, suggesting an origin of introns in a very early stage of evolution.

Archaeal Plasmids

In addition to the major circular chromosome, a number of *Archaea*, like many *Eubacteria*, also contain extrachromosomal genes, located on small circles of DNA called plasmids. These are particularly numerous in some of the halophilic *Archaea*, where they can carry a significant fraction of the cells' genetic information—for example, around 25 percent in the case of *Haloferax mediterranei*. Plasmids range widely in size, from tens to hundreds of kilobases, with large ones being particularly common in halophiles. Moreover, plasmids, unlike major chromosomes, often exist as multiple copies: cells of the thermophilic *Thermoplasma acidophilum* contain around ten copies of a 15 Kilobases plasmid called pTA1 and a variety of *Sulfolobus* has been observed to carry over forty copies of its 45 Kilobases plasmid, pNOB8. (Note that plasmids have been given catchy names, in which only the "p" is immediately explicable.)

It should be added that a plasmid is infective in the sense that it can be transferred from a cell that has one to another cell that doesn't. Plasmids can be absent from cells, and so clearly must carry genes that are not absolutely essential for life. As an example, many eubacterial plasmids carry genes that confer resistance to antibiotics; the cell-to-cell infectivity of such plasmids is one reason why antibiotic resistance is such a public health problem. Similarly, some thermophilic *Archaea* from ocean vents harbor plasmids that lead to heavy metal resistance, perhaps by encoding membrane-bound transport proteins that pump the metals from the cells. Such organisms include members of the genus *Alvinella*, which was named for the research submersible Alvin. These *Archaea* live in the body cavities of vent polychete worms.

Finally, other plasmids enhance the function of cells by unknown mechanisms: the sulfur-reducing thermophile *Desulfur-*

olobus ambivalens grows most rapidly when it contains a specific plasmid, pDL10. Unfortunately, one doesn't know what protein is encoded on that plasmid, so the mechanism for the enhancement of growth is unclear.

The relative autonomy of plasmids extends even further in the case of viruses. Viruses may be thought of as plasmids that carry information for the construction of a protein coat, and so can move from one cell to another in a more protected form. The *Archaea*, besides containing a good selection of plasmids, also include viruses* in their overall genetic systems. Viruses have been observed in all three categories of *Archaea*, the thermophiles, halophiles, and methanogens. Archaeal viruses invariably contain double-stranded DNA as genetic material, in contrast to eubacterial viruses, which sometimes employ either single-stranded DNA or even single-stranded RNA. The size of archaeal viral DNA ranges from less than 30 Kilobases (in sulfur-dependent thermophiles) to over 230 Kilobases in the halophilic *Halobacterium*.

Finally, the viruses associated with prokaryotes may be grouped into two categories: those that always destroy their host cell and those that may infect the host without necessarily harming it. The first are called lytic viruses because they invariably break down, or "lyse," their host. The second are temperate viruses: their temperate character is evident in their ability to go into a quiet state within the host, during which they become a gene on the host chromosome and can persist harmlessly in the host through many cell divisions. Both categories occur among the *Archaea*: at least two lytic viruses have been described in members of the genus *Methanobacterium*, and a temperate virus has been observed in *Methanolobus*.

Gene Transfer

Archaea and Eubacteria are noted for the variety of ways in which they can transfer genetic information between cells, ei-

---

*Viruses that infect microorganisms are often called phage, which is short for bacteriophage (meaning bacteria eater). Often, but not always, they are tadpole-shaped, and when such viruses infect their hosts, they attach via the tail and inject their genetic material into the cell through it.

ther within a species, or between members of different species. Mechanisms include specific transfer of plasmids, infection by temperate viruses, random uptake of DNA fragments from the environment, and transfer of the entire genome, or large parts thereof, through a mating process. Such "lateral" (or "horizontal") genetic transfer is exceedingly widespread in nature, and must be taken into account in thinking about the evolution of microorganisms and their impressive adaptability.

The study of intraspecies gene transfer is aided by the availability of mutant genes that can serve as "markers" and be followed through a transfer process. Therefore, it is useful that one has been able to induce mutations in a variety of thermophiles through mutagenic chemicals or radiation. These mutations, thus far, include genes affecting nutritional capability and sensitivity to inhibitory chemicals. These are easily assayed by inclusion (or omission) of the nutrient or inhibitor in the growth medium.

Transfer of marker genes between cells can occur by several mechanisms that have been demonstrated in *Archaea*. For example, transformation is the incorporation of external DNA into a cell, and can occur naturally (without intervention) or induced by laboratory procedures. Natural transformation has been demonstrated in several *Archaea*, but at a very low frequency. Cells of *Sulfolobus* and *Pyrococcus* have been rendered transformable—made permeable to external DNA—by treatment with calcium chloride, followed by a period of heating.

Sulfolobus acidocaldarius provides an example of the use of nutritional mutations in the study of transformation. First, several strains of this archaeon were isolated on the basis of their nutritional requirements, specifically those for amino acids. Then one strain that required one particular amino acid, but could synthesize all others, was mixed with a second strain that required a second amino acid, but was able to make the rest. Finally, the mixture was plated on solid medium that contained neither: only cells that contained hybrid DNA—cells that had been transformed by DNA transfer—would be able to grow. Such recombinant cells were detected in the form of colonies, indicating that gene transfer had occurred; in fact, it occurred at a rather high frequency. This experiment could be carried out successfully at temperatures as high as 84 degrees C and it has been suggested that an enhanced transformation ability may serve to counteract DNA instability owing to life at such high temperatures.

Other routes for gene transfer include transduction and conjugation. Transduction is the insertion of microbial genes by means of virus infection. The virus can be said to serve as a vector for the transferred genetic material. Thus far, transduction has been only demonstrated using a virus whose host is *Methanobacterium thermoautotrophicum.* And viruses have been identified only affecting a few sulfur-dependent thermophiles and methanogens, so that the possibilities for transduction seem quite limited. Conjugation is the transfer of genetic material through direct cell contact—a sort of microbial cell-to-cell mating. Examples of conjugation include transfer of genes between cells of the thermophile *Sulfolobus* as well as those of the halophile *Haloferax.* A number of different conjugation-transferred plasmids have been described in *Sulfolobus islandicus*, a close relative of *S. acidocaldarius.* These plasmids are efficiently transferred throughout a population, and can carry certain nonplasmid genes into a recipient that lacks them. It is a distinctive feature of such plasmids that they confer immunity to subsequent infection by the same kind: when such a plasmid enters a recipient cell, the cell becomes unable to receive additional copies.

An unexpected feature of microbial gene transfer has been the paucity of barriers that prevent interchange between different species. Indeed, eukaryotic species are defined in part by the rarity of interspecific genetic mixing: gene transfer usually occurs only within a species. But *Eubacteria* and *Archaea* carry out interspecific mixing with ease: genes move unhindered throughout the entire prokaryotic world, giving the concept of species a rather different meaning than in the eukaryotic case. We mentioned that such lateral gene transfer presents serious difficulties for the use of sequences in determining evolutionary relationships, and that rRNA sequence analysis was developed as one response to this problem.

Not only do genes move freely between remote species within the *Archaea* and *Eubacteria*, but it is now clear from sequence studies that genes move even between members of different domains, between the *Eubacteria* and the *Archaea*, and vice versa. It is as though all prokaryotic microorganisms somehow constitute a single "hyperspecies" with more or less free genetic transfer among members linking it together. The recent discovery that clusters of taxonomically unrelated microorganisms constitute metabolic units, requiring the proximity and contributions of the participants, supports the same idea: autonomous species may not be the natural "units" of microbiology. And if such clusters of microbes must cooperate in carrying out metabolic processes, then it follows that they require the presence of one another for viability. This requirement is believed by some to account for that fraction of microorganisms in nature that have not been grown individually in the laboratory.

Evidence for interspecific gene transfer in the *Archaea* is mostly indirect based on sequence studies. However, one direct example exists in species belonging to the halophile genus *Haloferax*. In these organisms, transfer of chromosomal DNA, as well as plasmids, can occur within a single species and even between the different species, *H. mediterranei* and *H. volcanii*. Unlike some eubacterial mating, the transfer occurs in either direction: there is no unique donor or recipient.

Finally, a discussion of gene transfer would be incomplete without mentioning evidence indicating that genes pass from individuals belonging to either prokaryotic domain to those belonging to the *Eukarya* (and, again, vice versa). Examples of such interdomain transfer include genes encoding the enzymes glyceraldehyde phosphate dehydrogenase, glucose phosphate isomerase, and superoxide dismutase. For example, there is a eubacterial glyceraldehyde dehydrogenase that is clearly eukaryotic in origin. *Escherichia coli* contains two versions of this enzyme, one similar in sequence to that of eukaryotes and one to that found in other *Eubacteria*. On the other hand, there is a superoxide dismutase in eukaryotic organisms that was evidently derived from a eubacterium—clearly transfer in the reverse direction.

Gene Expression: Archaeal RNA Polymerase

As we have already noted, protein synthesis begins with transcription—the synthesis of a messenger RNA (mRNA) molecule whose base sequence is complementary to, and determined by, that of the original gene. Messenger RNA synthesis is catalyzed by an enzyme complex called RNA polymerase, which binds to a region on the DNA called the promotor region and then moves along the DNA strand, reading (transcribing) the genetic message—that is, forming mRNA.

Structural studies of RNA polymerases indicate strong evolutionary ties between *Archaea* and *Eukarya*, but not extending to *Eubacteria*. RNA polymerase is an enzyme complex composed of a number of distinct protein molecules. The eubacterial complex is relatively simple, with only four distinct protein subunits. In contrast, both the archaeal RNA polymerase complex, and that of the *Eukarya*, are much larger and much more complicated. The complex in halophiles and methanogens contains eight individual subunits, while thermophiles contain ten or more. The corresponding RNA polymerase in eukaryotes also consists of ten or more subunits and these are roughly comparable in size to those of thermophiles. In general, the corresponding subunits in *Archaea* and *Eukarya* have similar amino acid sequences, although it should be said that a couple of the smaller subunits in *Archaea* bear no homology to those of any other organism. Most important, the sequences of subunits of both *Archaea* and *Eukarya* are strikingly unlike those of the *Eubacteria*.

Finally, there is another similarity between archaeal and eukaryotic transcription: the structure of promoter sequences in eukaryotic and archaeal cells is quite alike, and both are very different from those in *Eubacteria*. In the promoters, there are specific base sequences, not occurring in *Eubacteria*, but present in both of the other domains. It also turns out that the binding of archaeal or eukaryotic polymerase to the promotor region requires additional proteins, or "transcription factors." Significantly, these are not a part of eubacterial transcription. Thus, although we observed structural similarities between archaeal and eubacterial genomes, a number of features of transcription are quite distinct in the two groups. In such cases, the *Archaea* much more closely resemble eukaryotic organisms. How is this possible?

Gene Expression: Archaeal Protein Synthesis

The second half of gene expression is translation, with the protein's amino acid sequence being determined by the mRNA nucleotide base sequence. Recall that individual amino acids are positioned on the mRNA template (blueprint) by base pairing of adaptor molecules—the transfer RNAs, abbreviated tRNAs. Base sequences of these tRNAs are encoded in the DNA of specific tRNA genes. Thus, not all gene expression leads to protein synthesis: some genes specify sequences of the various transfer and ribosomal RNAs. While the role of tRNA is similar in all three domains, we saw that the archaeal and eubacterial tRNA genes are similar in containing introns, which *Eubacteria* lack. Again, note the similarity between *Archaea* and *Eukarya*. We must ask how, in light of what we know of the evolution of these groups, such a thing is possible.

Protein synthesis takes place at ribosomes, organelles composed of RNA and protein. The complexity of ribosomes is evident in their protein composition: although they contain only three or four types of rRNA, they are equipped with at least fifty-five different proteins. The ribosomes bind to messenger RNA molecules and then move along them, "reading" their sequence and synthesizing protein. We learned earlier that prokaryotic ribosomes—those of both *Archaea* and *Eubacteria*—are measurably smaller than those from eukaryotic cells, and the individual ribosomal RNAs are smaller, as well. Thus, prokaryotic ribosomes "weigh" 70S, the eukaryotic ones, 80S. Moreover, the two subunits composing the prokaryotic ribosomes weigh 50S and 30S, whereas the corresponding eukaryotic subunits weigh 60S and 40S. Finally, we learned that the 23S, 16S, and 5S rRNAs of prokaryotes correspond—that is, are related through evolution—to 28S, 18S, and 5S rRNAs in eukaryotes.

We wrote of "prokaryotic ribosomes" as if there were no distinction between those from the archaeal and eubacterial domains. However, when one looks closely at them, important differences emerge. Archaeal ribosomes are richer in protein than are their eubacterial counterparts and somewhat larger. Because protein is less dense than RNA, any difference in mass is largely obscured by the difference in protein content, and the two sorts of ribosome appear more similar than they really are. Thus, in regard to both ribosomal size and protein content, *Archaea* are intermediate between *Eubacteria* and the eukaryotes. And those *Archaea* believed to be the least evolved from common ancestors—organisms like *Thermococcus, Pyrococcus,* and *Archaeoglobus*—possess the largest and the most protein-rich ribosomes. In other words, they most closely resemble the eukaryotic situation.

Archaeal ribosomes also fall somewhere between those of the *Eubacteria* and *Eukarya* in their overall shape. Thus, eukaryotic ribosomes show characteristic bumps, not found on bacterial ones. To be more precise, the small (40S) subunit in *Eukarya* exhibits a beak-like bump that gives it the rather eerie appearance of a miniature bathtub duck. And the larger, 60S unit shows two bulges, separated by a groove. The corresponding archaeal pattern is intermediate, with a variable number of such features present. The sulfur-dependent thermophiles like *Sulfolobus*, *Thermoproteus*, and *Desulfurococcus* possess most of the eukaryal features, whereas methanogens and halophiles are more like the *Eubacteria* in having perhaps only one of the eukaryal protuberances. These features clearly reflect underlying molecular structure; they support an evolutionary closeness between the *Archaea* and *Eukarya*, also evident from other criteria.

This far-reaching distinction between sulfur-dependent thermophiles and the other *Archaea* is vivid enough to have encouraged James Lake to advance a controversial proposal. Based on a variety of (mostly sequence) evidence, he suggests that these sulfur-dependent organisms should be placed in a separate domain, the *Eocytes*, which means dawn cells. In Lake's scheme, the *Eocytes* are the closest neighbors to the *Eukarya*. This distinction between thermophilic *Eocytes* and the other *Archaea* is reminiscent of Woese's distinction between the thermophilic *Crenarchaeota* and the rest of the *Archaea*, the *Euryarchaeota*. But, in Lake's view, the bifurcation is much deeper than in Woese's scheme, separating domains rather than kingdoms. In other words, *"Eocyte"* represents a different tree: a different evolutionary history for the *Archaea*. It should be noted that these matters are subjects of lively discussion but that, at present, many feel that Lake has not advanced sufficient evidence to support such an extensive reworking of archaeal systematics.

Inhibitors of Protein Synthesis

Protein synthesis is complicated, with numerous individual steps and a large cast of characters. For that reason, we will consider only two, relatively uncomplicated, aspects of the situation, the first being the inhibitory action of diphtheria toxin on protein synthesis that we have already encountered. This toxic protein is released by certain strains of the bacterial pathogen, *Corynebacterium diphtheriae*. The protein can kill eukaryotic cells at extremely low concentrations—in some cases, one molecule per target cell seems to be enough to do the trick. Diphtheria toxin kills cells by inhibiting protein synthesis at the stage where amino acid chain elongation occurs. The toxin kills eukaryotic cells and the cells of all three archaeal groups. In contrast, eubacterial cells are totally immune.

On the other hand, there are inhibitors of chain elongation, like the antibiotic kirromycin, that inhibit only eubacterial protein synthesis, leaving both *Archaea* and *Eukarya* unscathed. In both cases, the *Archaea* line up solidly with the *Eukarya*. Certain other inhibitors of protein synthesis emphasize the distinction between *Crenarcheota* and *Euryarchaeota*, but the overriding impression is always that all *Archaea* most closely resemble the *Eukarya*. Again we must ask what evolutionary meaning can be attached to such archaeal eukaryotic similarities.

Other Eukaryotic Connections

Proteasomes provide additional evidence for evolutionary connections between certain *Archaea* and *Eukarya*, but not *Eubacteria*. Proteasomes are enzyme complexes catalyzing the digestion of cytoplasmic proteins. In many instances, the proteins are targeted for digestion because their synthesis had been defective, leading to structural imperfections. In other cases, the proteins are simply worn out, having been degraded by oxidation. And in still others, they are proteins that have been taken up by the cell, serving as dietary energy sources. These proteins are targeted for digestion by the covalent binding of a small protein called ubiquitin, which, as its name indicates, is widely distributed, occurring in all eukaryotic cells. However, ubiquitin has never been observed in a eubacterial organism. Significantly, ubiquitin, as well as proteasomes, have been observed in the archaeal genus, *Thermoplasma*. There, the proteasomal complex is much simpler than in eukaryotic organisms, consisting of two subunits, instead of the twelve or more occurring in *Eukarya*. A case has been made for this archaeal proteasome being ancestral, a precursor of the eukaryotic version. It is interesting that ubiquitin and proteasomes have not been observed in other *Archaea*, although serious efforts have been mounted. This failure either reflects special difficulties of detection or a special (and intriguing) connection between *Thermoplasma* and the eukaryotic domain. The plausibility (but not the nature) of such a connection is reinforced by the observation that *Thermoplasma* is almost unique among *Archaea* in containing histone-like proteins that bind to its DNA exactly in the manner of all eukaryotic organisms. (*Methanothermus*, a thermophilic methanogen not too distantly related to *Thermoplasma*, also has histone-like proteins.)

Finally, one must take into account the matter of heat shock proteins. These constitute a family of proteins that fall into two main categories. Some, the chaperonins, play a role in protein folding after synthesis, preventing new proteins from adopting erroneous three-dimensional shapes. Others are only synthesized when cells are subjected to excessive heat or other insults capable of affecting normal protein structure. These heat shock proteins exert a protective effect on cellular proteins and nucleic acids, stabilizing their three-dimensional structures. They are widespread in all three domains.

In contrast to evidence suggesting evolutionary relationships between *Archaea* and *Eukarya*, heat shock proteins indicate possible (and puzzling) affinities between *Eukarya* and certain kinds of *Eubacteria*. Thus, the gene for a particular heat shock protein, Hsp 70, has been sequenced in a number of organisms, including several thermophilic and halophilic *Archaea*. When sequences are compared, those from *Archaea* are not all that similar to those from the *Eukarya*, but instead more closely resemble the Gram-negative bacteria, one of the two major structural categories of the *Eubacteria*. On the other hand, by the same (Hsp) criteria, the *Eukarya* appear to be closer to the Gram-positive bacteria, the other major eubacterial category. We must ask ourselves how this is possible. Based on such evidence as specificity of inhibitors of protein synthesis, as well as ribosome structure, *Eukarya* appear evolutionarily connected to *Archaea*, whereas analysis of heat shock proteins yields a quite different picture. And what did Doolittle et al. mean, in evolutionary terms, by saying that archaeal genomes exhibit "eubacterial form and eukaryotic content"? Either *Eukarya* evolved from the Archaea (or some of the Archaea) or they didn't. Right?

The Origin of Eukaryotes: Endosymbiosis

Life began almost four billion years ago, but the earliest eukaryotic fossils are only a little more than one billion years old and it is likely that the *Eukarya* could not have arisen before about two billion years ago. So roughly the first half of evolutionary history was exclusively prokaryotic, with the Earth inhabited only by *Eubacteria* and *Archaea* or their joint precursor. Therefore, if the first important landmark in evolution was life's origin, and the second, the divergence of *Eubacteria* and *Archaea*, the third great event was certainly the emergence of the Eukarya.

Eukaryotic cells are more complex than anything that existed before. They contain a membrane-bounded nucleus, have characteristically larger ribosomes, and usually contain membrane-bounded organelles such as mitochondria (that couple ATP synthesis to oxidation of metabolites) and chloroplasts (that perform photosynthesis). Such organelles enjoy a measure of genetic autonomy within the cell: they contain genetic information of their own, located on a small, circular, prokaryote-style chromosome.

An important advance in our understanding of the rise of *Eukarya* was the "endosymbiont" hypothesis proposed* in the 1960s by Lynn Margulis. This powerful idea was that mitochondria, chloroplasts, and probably other eukaryotic structures arose by endosymbiosis of appropriate prokaryotic cells. Thus, host cells engulfed bacteria, which then persisted inside them in a symbiotic relationship, with the guest evolving in parallel with the host. The organelle-to-be received nutrients and protection from the host and provided it, in the two examples mentioned, with energy from cell respiration (mitochondria) or photosynthesis (chloroplasts).

It is evident that considerable genetic integration has occurred between host and guest organelle. Some mitochondrial proteins are encoded partly in mitochondrial DNA and partly in

---

*Actually, revived is a better word: the idea had been around during the nineteenth century, but with rather little evidence to support it.

the host nucleus and entire genes have been exchanged in both directions between host and endosymbiont. Moreover, endo-symbiotic events have occurred repeatedly during evolution, as chloroplasts and mitochondria undoubtedly arose separately, while, additionally, the chloroplasts of different algae probably arose in independent episodes. On the basis of rRNA sequence comparisons, among other evidence, it seems that chloroplasts originated as engulfed photosynthetic *Eubacteria*, probably evo-lutionary precursors of the cyanobacteria, and mitochondria as aerobic, Gram-negative *Eubacteria*. The last probably belonged to a group called purple bacteria, which include the modern-day symbiont *Rhizobium*, which lives in plant nodules and nourishes its host with nitrogen from the atmosphere. Many plant cells contain both mitochondria and chloroplasts; their evolutionary histories include at least two endosymbiotic occurrences.

Support for the endosymbiotic origin of organelles has in-cluded the study of very primitive *Eukarya* who descend from a line of ancestors that may predate the event. Thus, *Giardia*, the causative agent of "beaver fever" and *Pelomyxa*, a so-called giant amoeba, lack mitochondria. But, interestingly, *Pelomyxa* seems to be "caught in the act" of endosymbiosis because it contains several different intracellular microorganisms that ap-pear to assist in its metabolism. And, from our point of view, it is particularly interesting that these internal microbes include both *Eubacteria* and methanogenic *Archaea*.

The Origin of Eukaryotes: A Scenario

So it is likely that eukaryotic cells arose from multiple and in-dependent lineages by means of a succession of endosymbiotic fusions. And with endosymbiosis in mind, it is possible to con-sider the question of the evolutionary origin of eukaryotic cells and, in particular, the eukaryotic nucleus. And that question turns out to be closely connected to the matter of similarities between eukaryotic genomes, on the one hand, and those of both *Archaea* and *Eubacteria*, on the other—i.e., how is it pos-sible that the eukaryotic nucleus (and genome) have both ar-chaeal and eubacterial features? And, if there was indeed en-dosymbiosis in eukaryal history, what was the identity of the invader and the host organisms?

An interesting contribution to these issues was made by James Lake in 1994 when he wrote, with M. C. Rivera, a paper entitled "Was the nucleus the first endosymbiont?" Based partly on heat shock protein sequences, the authors suggested that the eukaryotic nucleus originated through an endosymbiosis event, in which the "host" cell was a Gram-negative eubacterium and the "guest," and progenitor of the nucleus, a thermophilic member of the *Archaea*. Lake had proposed naming these thermophilic organisms *Eocytes* based on his perception of their deeply branched nature. That name, denoting dawn, takes on special resonance if the *Eocytes* are truly ancestors of the eukaryotic nucleus. In the sense that the nucleus, chief repository of the cell's genetic information, exerts primary control over the entire cell, the archaeal guest clearly took over the whole business. And so, in that respect, it may have become the true ancestor of us all.

This hypothesis—the eukaryotic nucleus originating as an archaeal guest penetrating a eubacterial host—may account for connections between eukaryotic and archaeal genetic systems that have been described. And the eubacterial host may be the origin of eubacterial features that also are apparent. Of course, it is also necessary to invoke losses of eubacterial traits in the final eukaryotic product. Why, for example, aren't eukaryotic genes arranged in operons, as in both prokaryotic domains? But traits are often lost in the course of evolution, so perhaps this is not such a serious problem.

But another point requires mention: in the Lake scenario, the nucleus is archaeal; the cytoplasm is derived from the eubacterial host. Is this really necessary—did the nucleus have to arise through endosymbiosis? After all, the archaeal cell could have simply been the host in the events that led to incorporation of mitochondria and chloroplasts—the cytoplasm could be, in other words, mostly archaeal. Any eubacterial features could then have come through the well-established ability of prokaryotes to exchange bits of genetic information.

Such an explanation is certainly possible, but one additional observation promotes the idea of an endosymbiotic nucleus. The primitive, pre-mitochondrial eukaryote *Giardia lamblia* contains two different heat shock proteins that both occur generally in *Eukarya*. One of the two versions occurs free in the cytoplasm, the other in the endoplasmic reticulum, a system of double membranes that are continuous with the (double) nuclear envelope. The nuclear envelope, with the connected endoplasmic reticulum, probably arose from a combination of the plasma membranes of both partners in the endosymbiosis. This is analogous to the way that the outer mitochondrial membrane apparently came from the host and the inner one from the guest. It appears that both the cytoplasm and the endoplasmic reticulum versions of the heat shock protein are similar in sequence to a prokaryotic version found in Gram-negative *Eubacteria*, but not in *Archaea*. Thus, the host cell was presumably a eubacterium, as Lake's proposal requires.

There is much yet to be discovered about genome relationships between the *Eukarya* and possible prokaryotic progenitors. In reading about the topic, one more and more frequently encounters the word chimera. This word is apt: in classical mythology, a chimera was a monster, with parts assembled from a lion, a serpent, and a goat (and a chimera was thermophilic: it breathed fire) The eukaryotic cell may exhibit several levels of chimeric structure. Mitochondria and chloroplasts probably entered the scene as engulfed *Eubacteria*. The remainder of the cytoplasm may have originated with other *Eubacteria*, including the ones responsible for the endoplasmic reticulum heat shock protein. And, of course, Lake proposes that the nucleus was derived from an archaeal cell. Indeed, we learned that even the eukaryotic genome appears chimerical, with sequences derived from both *Archaea* and *Eubacteria*: this complexity should not be particularly surprising, given the multiple available sources as well as the great propensity of prokaryotic genomes for lateral gene transfer.

9

Looking Forward

We have known about the *Archaea* for roughly two decades, a period of rapid discovery and not a few surprises. During this time, archaeal biology has stimulated us to think a great deal about early evolutionary history, which began roughly 3.8 billion years ago. Somehow, our minds can accommodate such expanses of time with relative ease; looking back—even a long way back—seems to come naturally.

In contrast, looking forward only a few years can be a daunting matter, especially, for example, when scientists try to predict the immediate future in their own fields of endeavor. Viewed in retrospect, they seldom turn out to be very accurate; the unexpected is always the rule. And so it will undoubtedly be in archaeal biology, but here we will throw caution to the winds. So let us ask what problems will the next decade of archaeal research concentrate on, what insights may come from the next rounds of discoveries, and what scientific specialties will produce them? Unquestionably, more wonders are just around the corner. For that reason, the remainder of this chapter will consider a few hot topics that appear likely to drive events for at least a little while.

How We Arrived at the Present Situation

During the past twenty years, emphasis has understandably centered on the unique features of *Archaea* and especially their remarkable adaptations to extreme environments. Those archaeal attributes that appeared to break well-established rules came in for special attention. For example, the discovery that archaeal membranes were not necessarily phospholipid bilayers was a bombshell that stimulated a great deal of research. Likewise, the unprecedented thermal stability of many archaeal proteins encouraged research in protein structure and led to interesting practical applications as well. And, of course, the deep and unexpected evolutionary isolation of the *Archaea* as a whole simply stunned the scientific community. This evolutionary isolation was widely viewed as reflecting ecological isolation: spectacularly forbidding habitats protected the *Archaea* from the vicissitudes of natural selection. Indeed, in the first fifteen (or so) years that they were known, the *Archaea* were considered to inhabit only such hostile environments, and not normal ones. "Normal" in this context of course means suitable—or pleasant—for humans.

We were all dazzled by such wonders, such singular organisms, such apparent remoteness from the rest of life. But, most recently, there were paradoxical signs of apparent connectedness that began to attract interest. Thus, the *Archaea* were discovered, through the proxy of their rRNA signatures, to inhabit diverse and often friendly environments: garden soil, the sea, the interiors of other organisms. They were everywhere—not just in hellish places. And after detailed study of genetic organization and gene expression, they appeared to be *all-too-well* evolutionarily connected: as we saw previously, they resembled the true bacteria (*Eubacteria*) in some respects, the more complex eukaryotic organisms in others. Apparent evolutionary isolation contrasted with a complex set of relationships, some of them based on endosymbiosis, others perhaps on lateral gene transfer.

So, when thinking about the future course of archaeal study, we are tempted to predict that the ubiquitous distribution of these creatures, together with the complexity of their evolutionary relationships, will provide much of the research focus. And because some features of the evolutionary complexity appear related to the earlier participation of *Archaea* in endosymbiotic alliances, the occurrence of modern-day (archaeal) methanogens in the cytoplasm of eukaryotic organisms, such as the "giant amoeba" Pelomyxa, will receive close scrutiny.

There is also the matter of practicality. Nothing determines research focus quite as well as the profit motive, and commercial benefits are beginning to accrue from the availability of thermally stable enzymes from hot spring *Archaea* (and also *Eubacteria*). In addition, the formidable chemical capabilities of *Archaea* are increasingly likely to give them a place in the microbial armamentarium of chemical engineers and pharmaceutical chemists.

Extending the Scope of Archaeal Ecology

We have already noted our emerging ability to identify *Archaea* in diverse ecosystems based on their nucleotide signatures. Although many organisms identified in this fashion have never been cultured, this is not to say that they cannot be. One therefore predicts that the next several years will see vigorous efforts being made to obtain cultures of the various categories of *Archaea* from such nonhostile inocula. And archaeal rRNA signatures are not just found in soil. *Archaea* are turning out to be major players in oceanic ecosystems. Thus, they constitute a major part of the picoplankton—mostly prokaryotic, floating organisms in the micrometer range—living in relatively cold ocean water masses. Because of their global distribution and the large ocean volumes involved, these organisms are significant contributors to the world ecosystem and it is a great pity that we know so little about them. However, from their sequence signatures, it is possible to identify many of them as belonging to the methanogenic tribe, although the environment is hardly anaerobic and actual synthesis of methane exceedingly unlikely.

Other sequences have been identified as those of the sulfur-dependent thermophile group, the *Crenarchaeota*. This is a rather enigmatic discovery because the sequences often crop up where the water is cold. Clearly, the usual expectations don't apply and, for example, an organism with a thermophilic rRNA signature may not actually be thermophilic at all, but just genetically related to known organisms that are.

It is also interesting that the sequences cluster in groups: it appears that these planktonic *Archaea* live in communities of closely related organisms. It is possible that the clusters represent physiologically connected organisms whose proximity reflects contribution to a communal metabolic process, a phenomenon described in the previous chapter. But we really don't know if this is the case.

It is intriguing to learn that numerous uncultured microorganisms abound in just about any ecosystem that one can think of. Thus, microbiologists have been isolating *Archaea* and *Eubacteria* from Yellowstone hot springs for years, but rRNA sequence studies have disclosed an unexpected degree of diversity. A plethora of *Archaea* that have escaped being cultured are nonetheless present. This is particularly intriguing because such hot springs are not normally considered paragons of diversity or ecological complexity. Moreover it is likely that some microorganisms that we know currently only through their rRNA sequence alias will be isolated when just the right conditions are employed. Microbiologists necessarily base enrichment techniques on their expectations regarding the physiology of the target organisms and those expectations are progressively expanding.

But the diversity story extends beyond the mere discovery of previously uncultured organisms. Sequences are now being discovered that identify their organisms unambiguously as *Archaea*, but as belonging to no known subgroup of that tribe. In other words, there appear to be major archaeal categories—as major as the methanogens or the halophiles—whose members have not yet been cultured and about whose attributes we cannot even guess. Such "deeply branching" organisms have no known close relatives. They have been discovered in extensively "picked-over" hot springs in Yellowstone National Park as well as in the open ocean. One example of deeply branched *Archaea*, the *Korarchaeota*, consists of organisms with sequences that only remotely resemble extreme thermophiles, differing as much from them as animals do from plants.

We simply don't know how such organisms fit into the evolutionary scheme and have no idea about how they make a living. Given our present ignorance of the evolutionary relationships of some of these "new" organisms, it seems likely that major revisions of the archaeal tree, including its connection with the (global) tree of all life, will be required during the next few years. With our total ignorance of the physiological and metabolic strategies of these new organisms, it is equally likely that major discoveries are ahead. Indeed, one should not be too surprised to encounter totally new kinds of metabolism, modes of cellular energy transfer as different from previously known sorts as, say, photosynthesis is from methanogenesis. Clearly, the occurrence of these new organisms indicates that the archaeal story will continue to unfold for some time and that some of that unfolding will be spectacular.

Archaeal Troglodytes (Deep in the Earth's Crust)

In terms of total mass, the *Archaea* are major players in the world, and not members of a marginal microbial sideshow. Up to 30 percent of the very small plankton floating in the colder regions of the ocean consist of unknown *Archaea*. Unhappily, we know virtually nothing about the physiology of these organisms or the contributions that they make to the marine ecosystem. It is also plain that archaeal contributions to other ecosystems are similarly important and similarly little understood.

With this ubiquity in mind, we now consider the proposal—some would say the "notorious proposal"—recently advanced by Thomas Gold from Cornell University. He suggests that deep regions of the Earth's crust contain extensive reservoirs of hot water that are the home to unimaginably large populations of thermophilic microorganisms, such as *Crenarchaeota*, thermophilic methanogens, and thermophilic *Eubacteria*. Based on deep-drilling encounters with this population, Gold judges that the whole subsurface biomass may be equivalent to the total of all surface life, terrestrial and marine. Thus, he envisions a doubling in the known living component of the planet.

Clearly, organisms living in pockets of water deep in the Earth's crust obtain their energy and substance from processing geochemicals—for example, substances of geochemical origin. Indeed, their available energy source must be electron transfer reactions—oxidations and reductions—between such mineral components as sulfur, hydrogen, and iron, while a chief source of carbon must be carbon dioxide. Organisms that obtain energy from oxidation of inorganic nutrients are called lithotrophs. They do not ultimately depend on photosynthesis for energy, in the way that virtually all other organisms do.

Energy metabolism in these creatures is intimately linked to geochemistry: they live by means of chemical transformation of material, such as sulfide minerals or metal salts, produced by geological processes. In that sense, the real energy source of these organisms consists of the very processes that led to the formation of the earth in the first place.

Because the deep biosphere is physically remote from the rest of the living world, its organisms do not interact with the rest of life, either in a metabolic or a genetic sense. Thus, evolution in the two realms (above and below) would be expected to occur independently and, owing to the stability of the crust environment, the organisms occurring in the depths would be expected to evolve at a slow and regular rate. Thus, one can hope that these deep organisms provide a window on much earlier times, a unique sampling of very early evolution.

In fact, the only obvious natural communication from the deep organisms is in the form of hydrothermal vents, which might be said to offer samples of that deeper world on a continuous basis. Thus, vent microorganisms—or at least some of them—may be refugees from the depths and perhaps our simplest way of sampling the life there. In this connection, it is noteworthy that vents that are geographically remote from one another often contain the same varieties of *Archaea*, as if those hydrothermal waters were in communication with the same subsurface reservoirs. Of course, in the case of submarine vents, it is possible that similar populations result from one vent seeding the next via cells floating in the intervening seawater.

However, land-based hot springs that are remote from one another also exhibit similar population structures. In such instances, overland migration seems quite unlikely and seeding by way of the atmosphere not much better. Thus, the case for subterranean transport via extensive seams appears much more probable. For instance, Mount St. Helens, after erupting catastrophically in 1980, soon exhibited a new hot stream, issuing from just below its crater. This stream was a product of the eruption and came from a region that had been completely sterilized—at over 600 degrees. However, in about a year, the stream was observed to contain at least two sorts of thermophilic *Archaea*: a methanogen and a sulfur-dependent thermophile. One must wonder how they got there. Springs in Yellowstone Park contain similar organisms, but they are 550 miles away. The choice seems to be underground travel over a great distance or long-term survival in an underground reservoir, possibly since the last previous eruption of the mountain, in 1800. In either case, Gold's vision of a hot, subterranean eco-system appears germane to a satisfactory solution to the recol-onization mystery.

There is other evidence that the "hot, deep biosphere" is not a silly idea and that microbiologists (and everyone else) should take it seriously. For one thing, humans have recently been drilling deeper and deeper into the Earth's crust, either in the course of oil exploration or for scientific reasons. In the course of such drilling, one encounters microorganisms, both archaeal and eubacterial. Initially, surface contamination was invoked to account for the presence of microorganisms in drill-ing fluids, but because many of these creatures were strongly thermophilic, it became increasingly difficult to pass them off as surface dwellers. Indeed, organisms encountered during deep drilling are often not only thermophilic—growing at tem-peratures over 90 degrees—but also barophilic. This means that they require high pressures, often several hundred times as high as that of the atmosphere. Obviously, such organisms originated at great depths.

The occurrence of large masses of microorganisms at ex-treme depths in the crust accounts for certain prior observa-tions that were rather enigmatic. Deep drilling also turns up significant quantities of methane as well as a variety of larger hydrocarbons. Methane is produced by methanogens and probably indicates large populations of them. At the same depths at which the methane (and the methanogenic *Archaea*) are found, also occur *Eubacteria* that metabolize methane. Thus, a complete methane-based ecosystem seems to be at work.

It is also evident that microbial denizens of oil deposits are busily engaged in the chemical alteration of hydrocarbon de-posits. For example, sulfur-dependent thermophiles, using sul-fur and sulfate in lieu of oxygen as a respiratory electron ac-ceptor, account for much of the sulfide that contaminates oil deposits. (Such sulfide production is colorfully called the sour-ing of the deposit.) Oil itself is not immune to microbial deg-radation. Fluids from Alaska and North Sea wells turn out to harbor thermophilic *Archaea* and *Eubacteria* able to break down petroleum hydrocarbons in the absence of oxygen. Thus, they may contribute to modification of the hydrocarbon composition of an oil reservoir and may have a noticeable economic impact.

Of course, the methanogens among them are actually engaged in natural gas production, manufacturing methane, the commercially valuable hydrocarbon. Here, one might ask if microbes are responsible for production of one kind of hydrocarbon, what about others? In fact, currently there is serious interest in the notion that petroleum hydrocarbons are also products of microbial synthesis. This would imply that oil reserves could be, at least in principle, a renewable resource. Unfortunately, there is little real proof of this assertion, and few specific microorganisms have been identified as likely participants. Because petroleum deposits are likely to originate at high temperatures and pressures, it is clear that microbial candidates for oil making should be thermophilic and, most likely, given the methanogenesis precedent, thermophilic *Archaea*. Perhaps more information will become available as microbiologists increasingly probe the hot, deep biosphere.

Their Contribution to Human Enterprises

Investigators of the *Archaea* are often questioned about the utility of these organisms and, by implication, the value of the research they carry out. "What good are they, or what harm do they cause?" is the sort of question that conditions the level of much research funding. The fact is that such microorganisms do us relatively little "good" and even less "harm." Take, for example, the matter of disease. There is little likelihood that a creature hailing from concentrated brine or boiling spring water would find an appropriate home inside a human body. Thus, pathogenic thermophiles or halophiles should be regarded as extremely unlikely.

Methanogenic pathogens are at least a possibility: some methanogens inhabit the human gut and the relationship might conceivably take an ugly turn. The most obvious environmental restriction for methanogens is their anaerobic life habit, and anaerobic pathogens certainly exist among the Eubacteria. But perhaps the only hint of negative consequences of harboring methanogens is their possible contribution to the methane component of flatulence. However, one should still exercise caution in the matter of pathogenic organisms. Studies of rRNA have turned up archaeal organisms in all sorts of habitat: why not also in animal bodies? And there might be pathogens among such creatures. Certainly examples exist of eubacterial pathogens that were at one time believed to be entirely harmless. One should be alert for archaeal counterparts.

But if archaeal organisms cannot (yet) be counted as pathogens, they do sometimes affect human activities in other ways, but usually in a rather mild fashion. Certain halophiles can cause discoloration of salted fish, meats, or of sea salt that had been prepared by evaporation of seawater. In these cases, contamination affects only appearance, perhaps making the products more difficult to sell, but is not known to cause illness. Apart from the sulfide production in petroleum deposits that we mentioned earlier, there are few other examples of *Archaea* inflicting harm on human enterprise. No doubt, as we get to know these organisms better, other examples will emerge, but one imagines that the effects will be mild—or else we would probably have already experienced them.

Industrious *Archaea*

On the beneficial side of the ledger, one should point first to the increasing usefulness of heat-stable enzymes synthesized by hyperthermophilic *Archaea* and other thermophilic microorganisms. Enzymes are effective catalysts that have found extensive use in food processing, industrial production of chemicals, as well as many laboratory procedures in molecular biology. In addition, enzymes are widely employed in the home as additives to detergents and stain removers. But the use of most enzymes entails a serious downside: at the very temperatures favoring the reactions being catalyzed, most enzymes are destroyed through a process called heat denaturation. This amounts to a loss of catalytic activity due to alteration of the protein's three-dimensional structure. However, many enzyme applications are most effective at an elevated temperature, as in the case of clothes washing, many industrial syntheses, or, in the laboratory, in the familiar polymerase chain reaction (PCR).

Luckily, enzymes obtained from thermophilic organisms are frequently resistant to heat denaturation and can therefore be used at much higher temperatures. Such thermostable enzymes play a rapidly growing role in a number of industrial chemical processes. Examples include the transformation of sugars in the production of corn syrup, other sugar alterations connected with brewing, and reactions connected with the manufacture of detergents. For instance, the conversion of the common sugar glucose to its isomer fructose (or fruit sugar) is best carried out at 100 degrees; this is possible on a commercial scale using the thermostable enzyme glucose isomerase from the eubacterium *Thermotoga maritima.* (Although not archaeal, this thermophile appears to be one of the most ancient bacteria, closest to the point where the *Archaea* and *Eubacteria* diverged.)

Recently, an application for an archaeal thermostable enzyme surfaced in the somewhat unexpected area of fuel production. It has been suggested that two archaeal enzymes could form the basis for an industrial-scale production of hydrogen from starch or cellulose. (Hydrogen is the quintessentially clean fuel—its combustion product is water.) Starch and cellulose are sugar polymers obtainable from any plant material, including trees. The two enzyme in question are glucose dehydrogenase, which oxidizes glucose, and hydrogenate, which produces hydrogen gas. For reasons previously mentioned, it makes sense that both these enzymes are obtained from thermophilic organisms and, in fact, the source of glucose dehydrogenase is the archaeal *Thermoplasma acidophilum,* a thermophile, with optimal growth at about 70 degrees. The source of hydrogenase is the still-more thermophilic *Pyrococcus furiosis,* which lives up to its name with optimal growth at about 100 degrees.

The first step in the process is the hydrolysis of starch or cellulose to form the sugar glucose. Glucose dehydrogenase then oxidizes the glucose, passing electrons to the coenzyme, NADP. This coenzyme carries electrons between the two enzymes, so that the oxidation of glucose can be coupled to the reduction carried out by the hydrogenase, leading to the hydrogen production. The coenzyme is therefore a shuttle, reduced in the glucose dehydrogenase reaction, then oxidized by the hydrogenase one, then returning to be reduced by glucose again, and so on. This process is particularly interesting from an environmental point of view as it is a case of fuel production with a renewable starting material and without any waste gases. A by-product of glucose oxidation is gluconic acid, which is, however, a valuable industrial chemical. Even the ultimate burning of hydrogen is uniquely clean, yielding only carbon dioxide and water.

Exploitation of thermostable enzymes requires a fairly long period for research and development and, not surprisingly, most examples thus far employ enzymes from *Eubacteria*. However, inasmuch as the records for extreme thermophilic growth are held (hands down) by members of the *Archaea*, there seems little doubt about their role in the future. Extensive use of enzymes from archaeal sources probably awaits additional knowledge about the organisms themselves. In the meanwhile, a helpful shortcut has been the isolation of the genes encoding thermophile enzymes and their subsequent insertion, often via a plasmid, into a host organism such as *Escherichia coli*. This is a bacterium that we know very well how to grow and manage.

The ability to move genes from organism to organism suggests an additional potential contribution by the *Archaea*, one that may, in the future, eclipse all others. The *Archaea* constitute a rich, new source of available genetic diversity, whole new families of genes, many, as we will see, with no counterpart in bacteria or eukaryotes. It is not just that some archaeal genes encode particularly stable proteins, or proteins with other special features. Archaeal genetic material has, quite suddenly, added a vast new source of meaningful nucleotide sequences to the library of all sequences, the armamentarium of molecular biology. The plethora of new archaeal genes will undoubtedly turn up applications, but in view of the new techniques for sequences manipulation, so will fragments of genes. One can be quite sure that the discovery of "the third form of life" will prove to have added enormously to known biodiversity in both the traditional (numbers of species) sense, and the potentially more fruitful informational sense.

Cleaning Up Messes

Many *Archaea* are lithotrophic, obtaining energy by oxidation of inorganic (mineral) nutrients. Such organisms can alter their mineral environment, an ability now being exploited. We encountered cells that live on the surface of minerals such as iron pyrite; these, through their metabolic activity, chemically transform the minerals. Such microbial activity is potentially useful in, for example, the biological processing of mine wastes. Although *Eubacteria* have been the major players in such remediation efforts, the archaeal role is definitely increasing. For example, the *Sulfolobus* shows promise by virtue of its ability to oxidize sulfur as well as its related knack for dissolving iron pyrite, an iron sulfur mineral.

Thermoplasma is an inhabitant of coal mine refuse piles and may offer some possibilities for cleaning them up. We have already encountered this rather bizarre thermophilic organism in the context of its variable shapes, its lack of a cell wall, and its small genome. *Thermoplasma* has been most commonly isolated from coal mine refuse, where it metabolizes organic chemicals, including sulfur compounds. Thus, the organism lives by "processing" coal wastes and it seems that a promising strategy may be to include it in consortia of microorganisms that can, together, mitigate some of the more unfortunate results of mining activities.

So it appears that, when archaeal biologists are asked about the practical value of their work, they can only point to modest benefits and usually emphasize the importance of fundamental biological understanding. The *Archaea* fail to make us sick, and it is possible that they will also not contribute much to gross national products (although one may want to keep an open mind about this). However, the *Archaea* are undoubtedly making us rethink central issues connected with cellular life: its earliest history, the environmental extremes that it can endure, even its role in shaping the biosphere. One imagines that this may be benefit enough.

Archaeal Genome Projects

August 23, 1996, will prove a date to remember in archaeal history and, indeed, in the history of all prokaryotic biology.

On that day, the journal *Science* published a paper with the title "Complete genome sequence of the methanogenic archaeon, *Methanococcus jannaschii*: insights into the origins of cellular life." The paper describes research that had clearly been a team effort—quintessentially big science. The first author was Carol Bult, who led the sequencing team, and her name was followed by no less than thirty-eight others. The name of Carl Woese was in there (number thirty-seven) as were those of a number of other capable molecular biologists. The final name on the list was that of J. Craig Venter, a major figure in molecular genome mapping. The paper came from the Institute for Genomic Research (Rockville, Maryland), which enjoys the attractive acronym TIGR, and of which Dr. Venter was director.

The paper describes the first global nucleotide sequence— the total genome—for a member of the *Archaea*. Thus, it presented the nucleotide sequence of each gene in the organism and the overall order of those individual genes: a detailed map of the whole business. As an undertaking, it is comparable to the Human Genome Project, about which there has been extensive publicity. Of course the *Methanococcus* genome is much smaller than the human genome, but has already yielded a kind of information that cannot be forthcoming from the human counterpart. For instance, it provides strong evidence for a closer evolutionary relationship between the *Archaea*; and the *Eukarya* than with the *Eubacteria*.

The TIGR group has been an innovator in genome sequencing: Dr. Venter developed the controversial "shotgun sequencing" approach, in which an entire genome is fragmented at random and the fragments partially sequenced. The integrated genome is then deduced from overlapping fragments using enormous computer routines and additional sequencing done only as required to make sense of the whole. For instance, in the case of *Methanococcus*, over 36,000 fragments were formed and studied in producing the final map. This approach has turned out to be more rapid than the classical, and more orderly, procedure of marching through the genome gene by gene, Armed with Venter's newer approach, the TIGR group had already published two complete eubacterial sequences, those of *Haemophilus influenzae* and *Mycoplasma genitalium*. Both are obviously pathogenic organisms and were chosen to be sequenced for that reason. On the other hand, the genome of

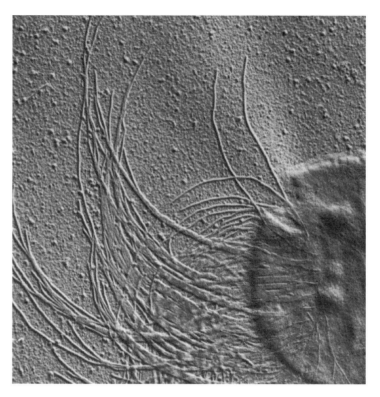

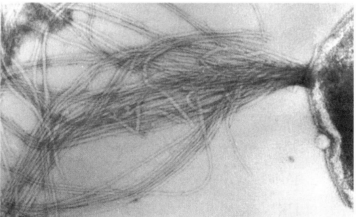

Figure 9.1 Electron micrograph of *Methanococcus jannaschii*, showing that it bears flagella on one side. (Courtesy of Professor John A. Leigh.)

Methanococcus jannaschii was chiefly interesting in an evolutionary sense. The organism was originally isolated from the base of a marine hydrothermal vent located in the Pacific Ocean at a depth of 2600 meters. It can grow at pressures in excess of 500 atmospheres and at a temperature of over 90 degrees. As its name indicates, it is able to produce methane, clearly both a methanogen and a thermophile. Its place in the archaeal tree, as determined by rRNA sequence analysis, is on the methanogen branch closest to the thermophiles.

The *M. jannaschii* genome consists of a main circular chromosome with 1,664,976 base pairs, a large circular extrachromosomal element (or plastid) with 58,407 base pairs, and a small circular plastid with 16,550 base pairs. In terms of genes, the main chromosome was estimated to contain about 1700, the large plasmid, about 44, and the small plasmid, only about 14. Regions of the chromosome could be identified as genes by looking for the signals that initiate and terminate the process of gene expression for each individual gene. To identify the genes as to function, it becomes necessary to compare, using a computer, the putative gene sequences with those of known genes— genes whose gene products have been identified in other organisms. Such information is found in huge databases, available for searching by computer networks. These contain sequences associated with identified gene products as well as those not yet associated with any known function. When this kind of searching was done, only a bit over one-third of the *Methanococcus* genes could be identified according to function. So almost two-thirds of the *Methanococcus jannaschii* genome remains unidentified, genetic information with an unknown role. Finally, roughly an additional quarter of the apparent genes corresponds to sequences that have been detected in other organisms but whose function has not yet been established.

Let us put it the other way around: large regions of this genome, in fact about 40 percent of the total, consist of sequences that have no clear counterpart in any other organism that has been studied. By comparison, about 80 percent of genes in *Haemophilus influenzae* corresponded to already known sequences and almost 60 percent matched genes of known function. These observations suggest that we have an enormous amount to learn about the physiology and molecular biology of *Methanococcus*, certainly a prospect to cheer the heart of any

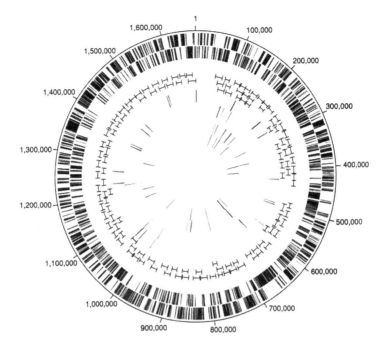

Figure 9.2 The main chromosome of *Methanococcus jannaschii.* This overview illustrates its circular character. Base pairs are numbered in the clockwise direction; individual lines represent regions that encode individual proteins. Because the chromosomal DNA is double-stranded, information can be carried on either strand. The outer concentric ring represents proteins encoded on the "plus" strand; the next, inner ring includes proteins encoded on the "minus" strand. (Additional inner rings reflect various special types of genetic information.)

curious molecular biologist. For example, it now becomes possible to go through the catalog of unknown genes and chemically knock them out one at a time, looking for lost functions. It is likely that completely new categories of metabolism will be discovered in this fashion, and that novel applications may well emerge. But the large question is this: what meaning can be attached to the "ordinariness" of the *Haemophilus* genome, in contrast to the relative remoteness of that of *Methanococcus*, with its lack of gene similarity to other organisms? More food for thought.

With regard to particular proteins, the TIGR study reported that many enzymes one would anticipate finding in this organism were indeed there. But some widespread enzymes that one might expect to encounter in any organism were notably absent. For example, the expected enzymes responsible for methanogenesis were present, as were enzymes responsible for carbon dioxide incorporation into organic molecules. On the other hand, a few enzymes catalyzing the synthesis of sugars from smaller precursors were present, while more were apparently absent. Does this mean that sugar synthesis in *M. jannaschii* is carried out by a different route from that of other organisms, or that the "missing" enzymes are actually there, but are based on sequences unlike those from any other sources? At present, one can't say which.

A variety of other genetic structures were observed in this genome. There were, for instance, regions that appeared to represent introns (intervening sequences), regions within a gene that do not encode proteins and are physically removed at the stage of messenger RNA. Recall that introns are not a usual feature of eubacterial genomes. Also observed were repeating sequences, a feature of virtually all genomes. And there was indirect evidence of the presence of transposons—the famous "jumping genes." Evidence of these consisted of sequences identified as encoding transposases, enzymes that catalyze the "jumping" process.

It should be said that other teams of scientists are industriously attacking the genomes of other organisms, including a number of other *Eubacteria* and *Archaea*. In fact, the results of completed genome projects are being published with increasingly frequency. For instance, such *Eubacteria* as *Escherichia coli*, *Bacillus subtilis*, and *Helicobacter pylori* have now been completely sequenced, as have the *Archaea, Methanobacterium thermoautotrophicum* (a thermophilic methanogen), and *Archaeoglobus fulgidus* (a sulfur-dependent thermophile). And the much more complex genome of the eukaryotic microorganism *Saccharomyces cerevisiae* (a yeast) has been published in its about fourteen million base-pair entirety.*

---

*Although a great deal larger than the prokaryotic genomes that we have been considering, this yeast genome is still only about one-two hundredth the size of its human counterpart.

Additional archaeal genomes are well on their ways to being sequenced. For instance, W. Ford Doolittle from Dalhousie University in Halifax, Nova Scotia, and several associates from laboratories in Canada and the United States are quite far along on a *Sulfalobus solfataricus* genome sequencing project, and have published preliminary accounts of what they are finding. This organism, first encountered in Chapter 1, is the most extensively studied example of the *Crenarchaeota* kingdom of the *Archaea*. A great deal is known about its biology and biochemistry, so that knowledge about its genome can emerge in a clear context. It is probably too early to say much about the *Sulfolobus* genome, except to note that its genes seem to be arranged in clusters, with enzymes associated with related functions being grouped together. Such an arrangement is commonly observed in the other prokaryotic genomes and probably facilitates coordinated regulation.

What have these genome projects revealed to us about the place of *Methanococcus jannaschii,* and the other *Archaea,* in the evolutionary scheme of things? Well, perhaps three enormously important things. First, archaeal organisms seem evolutionarily isolated from the remainder of the living world. The discovery of a large fraction of archaeal genes that appear to be without homology with those in other organisms was, in fact, quite a bombshell. *Archaea* are really isolated in the genetic, and therefore evolutionary, sense. The rather striking dissimilarity between archaeal and eubacterial genomes emphasizes the evolutionary distance between the two prokaryotic domains. We are quite right to have abandoned the term *Archaebacteria.* On the other hand, it must be remembered that the evolutionary isolation is not total: we saw that many archaeal genes associated with metabolism resemble those found in *Eubacteria.* Indeed, it is possible that some of them may have originated in bacteria and been transferred to *Archaea* by some "lateral" process, as described earlier.

At the same time, these genome investigations have emphasized a surprising degree of connectedness between the *Archaea* and *Eukarya*—in particular, connections most evident in those genes that are associated with replication and expression of genetic information. Surely, on the basis of such similarities, the two can appropriately be called sister domains, with the sisterhood perhaps reflecting the mosaic character of eukaryotic cells and possible archaeal contributions to the eukaryotic nucleus.

Finally, this archaeal genome project, with its detailed inventory of enzymes, can open a vista overlooking the world of very early organisms. Such a wealth of information from one of the most deeply branched organisms available presents new opportunities for extrapolation. What sort of metabolism characterized the most recent common ancestor of all life? One notes that *Methanococcus, Methanobacterium,* and *Archaeoglobus* are capable of an autotrophic, carbon-dioxide-using, lifestyle. So too are the most ancient among the *Eubacteria.* Because these two prokaryotic domains are so remote from one another, it is difficult to imagine that autotrophy is not a primitive characteristic. Therefore, it seems very likely that the last common ancestor was a user of carbon dioxide. The ancestor probably wasn't methanogenic, a capability that appeared to evolve only in the archaeal lineage. But the commonly held notion of the last common ancestor being a denizen of a rich, primordial soup, making its living by metabolizing preformed organic compounds, now seems quite untenable.

As we continue to discover the identities of genes from this, and other deeply branched organisms, we will certainly come to know more, ask more refined questions, and, in that fashion, piece together other attributes of early cellular life. Which of the known metabolic pathways was responsible for carbon dioxide incorporation in ancestral cells? How did these pathways connect to other synthetic processes? At what point did photosynthesis originate and how did it become integrated, as it is now, with autotrophy? And when, in the course of evolution, did all this happen?

Isn't that always the way? One learns things that are really interesting, as in the results of these genome projects, and a hoard of new questions emerge. Isn't there any end to the process?

Fortunately, not.

Additional Reading

BOOKS

Bock, G. R. and Goode, J. A., Editors. *Evolution of Hydrothermal Ecosystems on Earth (and Mars?)*. Ciba Foundation Symposium 202, John Wiley & Sons, New York, 1996.

Brock, T. D., Editor. *Thermophiles. General, Molecular and Applied Microbiology*. John Wiley & Sons, New York, 1986.

Cone, J. *Fire Under the Sea*. William Morrow, New York, 1991.

Deamer, D. W. and Fleischaker, G. R. *Origins of Life*. The Central Concepts. Jones and Bartlett, Boston, 1994.

deDuve, C. *Blueprint for a Cell. The Nature and Origin of Life*. Neil Patterson Publishers, Burlington, NC, 1991.

Dixon, B. *Power Unseen—How Microbes Rule the World*. W. H. Freeman & Co., New York, 1994.

Gold, Thomas. *The Deep Hot Biosphere*. Springer Verlag, New York, 1999.

Gross, Michael. *Life on the Edge—Amazing Creatures Thriving in Extreme Environments*. Plenum, New York, 1998.

Horikoshi, Koki and Grant, William D., Editors. *Extremophiles—Microbial Life in Extreme Environments*. Wiley-Liss, New York, 1998.

Javor, Barbara. *Hypersaline Environments—Microbiology and Biogeochemistry*. Springer-Verlag, Berlin, 1989.

Kates, M., Kushner, D. J., and Matheson, A. T., Editors. *The Biochemistry of Archaea (Archaebacteria)*. Elsevier Science Publishers, Amsterdam, 1993.

Kluyver, A. J. and van Niel, C. B. *The Microbe's Contribution to Biology*. Harvard University Press, Cambridge, 1956.

Madigan, M. T., Martinko, J. M., and Parker, J. *Brock Biology of Micro-organisms*, eighth edition. Prentice-Hall, Upper Saddle River, NJ, 1997. (The latest edition of the authoritative textbook, of which T. D. Brock was senior author of the first seven. Chapter 17 is devoted to the *Archaea.*)

Margulis, Lynn. *Origin of Eukaryotic Cells.* Yale University Press, New Haven, 1970.

Morowitz, H. J. *Beginnings of Cellular Life.* Yale University Press, New Haven, 1992.

Parson, L. M., Walker, C. L., and Dixon, D. R., Editors. *Hydrothermal Vents and Processes.* (Geological Society Special Publication Number 87.) The Geological Society, London, 1995.

Postgate, J. *The Outer Reaches of Life.* Cambridge University Press, Cambridge, 1994.

Sonea, S. and Panisset, M. *A New Bacteriology.* Jones and Bartlett, Boston, 1983.

Staley, James T., Editor. *Bergey's Manual of Systematic Bacteriology*, Volume 3. Williams and Wilkins, Baltimore, 1989. (This is the volume that includes 80 pages devoted to the *Archaea*—here called *Archaeobacteria.*)

Woese, Carl R. and Wolfe, Ralph S. *The Bacteria—A Treatise on Structure and Function.* Volume VIII, *The Archaebacteria.* Academic Press, New York, 1985.

PERIODICALS

Brock, T. D. *"The Road to Yellowstone—and Beyond." Annual Review of Microbiology* 49, 1–28, 1995.

Bult, C. J. et al. "Complete genome sequence of the methanogenic archaeon, *Methanococcus jannaschii.*" *Science* 273, 1058–1072, 1996.

Edmond, J. M. and von Damm, K. "Hot springs on the ocean floor." *Scientific American* 2488, 78–93, 1983.

Fox, G. E., Magrum, L. J., Balch, W. E., Wolfe, R. S., and Woese, C. R. "Classification of methanogenic bacteria by 16s ribosomal RNA characterization." *Proc. Natl. Acad. Sci. USA* 74, 4537–4541, 1977.

Lake, J. A. and Rivera, M. C. "Was the nucleus the first endosymbiont?" *Proc. Natl. Acad. Sci. USA* 91, 2880–2881, 1994.

Madigan, M. T. and Marrs, B. L. "Extremophiles." *Scientific American*, April, 82–87, 1997.

Miller, S. L. "Production of amino acids under possible primitive earth conditions." *Science* 117, 528–529, 1953.

Schobert, B. and Lanyi, J. "Halorhodopsin is a light-driven chloride pump." *Journal of Biological Chemistry* 257, 10306–100313, 1986.

Walsby, A. E. "A square bacterium." *Nature* 283, 69, 1980.

Woese, C. R. "Archaebacteria." *Scientific American,* June, 107–122, 1981.

Woese, C. R. and Fox, G. E. "Phylogenetic structure of the prokaryotic domain: The primary kingdoms" *Proc. Natl. Acad. Sci. USA* 74, 5088–5090, 1977.

Wolfe, R. S. "My kind of biology." *Annual Review of Microbiology* 45, 1–35, 1991.

Index